Apollo 10

The NASA Mission Reports

Compiled from the NASA archives & Edited
by Robert Godwin

All rights reserved under article two of the Berne Copyright Convention (1971).
We acknowledge the financial support of the Government of Canada through the
Book Publishing Industry Development Program for our publishing activities.
Published by Apogee Books an imprint of Collector's Guide Publishing Inc., Box 62034, Burlington, Ontario, Canada, L7R 4K2
Printed and bound in Canada by Webcom Ltd of Toronto
Apollo 10 — The NASA Mission Reports
Edited by Robert Godwin
ISBN 1-896522-52-1
All photos courtesy of NASA

Introduction

Apollo 10 was a demonstration of excellence. By May of 1969 the United States National Aeronautics and Space Administration was prepared to send a second crew of three men into deep space. This time the trip to lunar orbit was to be taken by an all veteran crew consisting of Commander Tom Stafford, Command Module Pilot John Young and Lunar Module Pilot Gene Cernan. All three men had already acquired substantial amounts of flight time in space but once again the risk of traveling a quarter of a million miles away from Earth was paramount in everyone's minds.

The flight of Apollo 10 was still considered a test-flight. The Saturn V launch vehicle had undergone a series of modifications since Apollo 9 to make it capable of carrying a heavier payload and the Lunar Module's tracking and communication equipment had still not yet been tested over lunar distances. There were still many unanswered questions.

Commander Tom Stafford had flown on two previous Gemini flights during which he had successfully accomplished rendezvous with other space vehicles. During Gemini 7/6 Stafford took part in the first ever rendezvous in space and during Gemini 9 he accomplished three more rendezvous' with an unmanned target vehicle. Apollo 10 was to be his fifth space rendezvous.

Stafford's crew consisted of his Gemini 9 crewmate Gene Cernan and a veteran of Gemini 3 and 10 — John Young. Both Young and Cernan would later walk on the moon and Young would even go on to Command the first flight of the Space Shuttle. Never before had NASA assembled a more experienced crew.

Apollo 10 was to follow the entire flight-plan proposed for a lunar landing but would return after descending to only 47,000 feet above the lunar surface. Everything was to be the same; including the time of day and sun angles over the proposed landing site.

During the flight of Apollo 8, the previous December, it had been determined that the moon's gravitational field affected the space-vehicle differently depending on the altitude and course. One of Apollo 10's goals was to fly an identical mission trajectory as the upcoming Apollo 11 flight, to determine precise details of these gravitational fluctuations. It had been shown that the ground telemetry could be off by as much as 5,500 feet and so this "dry-run" was imperative.

Once the two space-craft separated in lunar orbit Cernan and Stafford descended to about 65,000 feet at which point the landing radar began to detect the lunar surface for the first time and sent it's telemetry back to Mission Control in Houston. Just prior to reaching their target altitude the descent stage of Lunar Module "Snoopy" was jettisoned and after descending for another ten minutes the ascent engine was ignited and the journey back to orbit began. It must have been simultaneously exhilarating and frustrating — so near, yet so far.

An oversight by the crew, however, had left the AGS system in Automatic rather than Altitude Hold Mode which led to the spacecraft beginning an unplanned maneuver, momentarily shaking the crew and vehicle. Voicing a few uncensored expletives Stafford and Cernan wrestled the vehicle back under control and began their ascent back to the waiting Command Module "Charlie Brown".

Although critics on Earth condemned the astronauts' use of so-called "street language" some found it refreshing and exciting to hear a crew of seasoned veterans commenting on their flight with such unbridled enthusiasm. Everything from the thundering lift-off, to Earth-rise above the moon, to the barren and battered lunar surface brought exultant comments from the three men.

To stress the test-flight nature of Apollo 10 the post-flight summaries included in the following pages show that nearly fifty discrepancies showed up in the equipment during the flight, but considering the millions of pieces which comprised the Apollo-Saturn space vehicle this was an incredibly small number.

During the course of compiling the data for these books it became apparent that each crew used their available film footage differently. Apollo 10 was the first to send color television pictures back from the moon but the amount of color still pictures shot on the 70 mm camera was a small portion of the images shot. One of the mission goals was to take as many pictures as possible of the potential landing sites on the moon's surface. Included on the enclosed CD-ROM is the entire run of 70mm still photographs taken by the Apollo 10 crew. It consists of over 1300 shots which are almost all of the lunar surface with very few candid shots taken by the crew of themselves. Some beautiful color imagery was acquired of the two space-craft in lunar orbit and these are also included on the CD-ROM although due to space limitations all of the pictures are in low resolution.

The planning and execution of the flight of Apollo 10 was so perfect that five of the seven mid-course corrections were cancelled. Due to the almost flawless flight, which resulted in the crew splashing down a mere three miles from their designated target, Apollo 10 has assumed an inconspicuous place in the history of America's voyages to the moon. It should however be remembered that every flight of the Apollo space vehicle was unique and required different hardware to be built and different training to use it. While John Young and Gene Cernan would go on to walk on the moon during the flights of Apollo 16 and 17 and would ensure their places in the history books by being the first to fly the Space Shuttle and the last to walk on the moon — it was Tom Stafford's skill in rendezvous and navigation during the flight of Apollo 10 that first put the Lunar Module through its paces in its intended environment — lunar orbit.

The final piece of the immense puzzle was now in place — ready for the flight of Apollo 11.

Robert Godwin
(Editor)

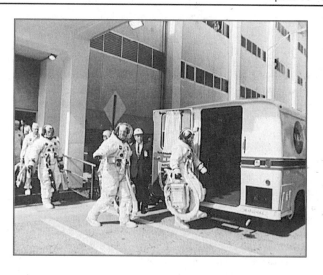

Contents

APOLLO 10 PRESS KIT

MISSION OPERATION REPORT APOLLO 10 (AS-505) MISSION

LIST OF FIGURES

LIST OF TABLES

POST LAUNCH MISSION OPERATION REPORT

LIST OF TABLES

POST LAUNCH MISSION REPORT NO. 2

LIST OF TABLES

PRESS KIT

FOR RELEASE: Immediate

May 7 1969

RELEASE NO: 69-68

PROJECT:APOLLO 10

APOLLO 10: MAN'S NEAREST LUNAR APPROACH

Two Apollo 10 astronauts will descend to within eight nautical miles of the Moon's surface, the closest man has ever been to another celestial body.

A dress rehearsal for the first manned lunar landing, Apollo 10 is scheduled for launch May 18 at 12:49 p.m. EDT from the National Aeronautics and Space Administration's Kennedy Space Center, Fla.

The eight-day, lunar orbit mission will mark the first time the complete Apollo spacecraft has operated around the moon and the second manned flight for the lunar module.

Following closely the time line and trajectory to be flown on Apollo 11, Apollo 10 will include an eight-hour sequence of lunar module (LM) undocked activities during which the commander and LM pilot will descend to within eight nautical miles of the lunar surface and later rejoin the command/service module (CSM) in a 60-nautical-mile circular orbit.

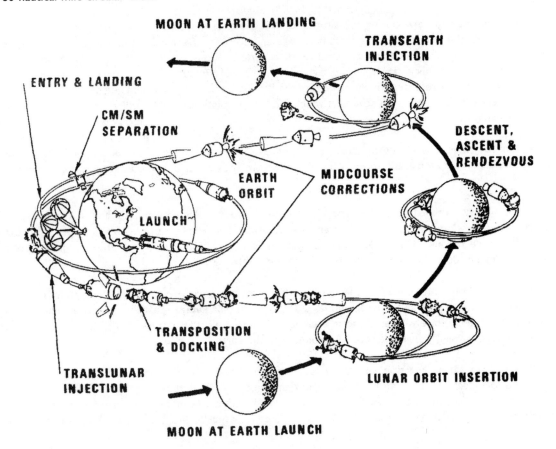

APOLLO LUNAR MISSION

All aspects of Apollo 10 will duplicate conditions of the lunar landing mission as closely as possible — Sun angles at Apollo Site 2, the out-and-back flight path to the Moon, and the time line of mission events. Apollo 10 differs from Apollo 11 in that no landing will be made on the Moon's surface.

Apollo 10 is designed to provide additional operational experience for the crew; space vehicle; and mission-support facilities during a simulated lunar landing mission. Among desired data points to be gained by Apollo 10 are LM systems operations at lunar distances as well as overall mission operational experience. The LM was successfully checked out in Earth orbit in Apollo 9, including a rendezvous sequence simulating lunar orbit rendezvous.

Space navigation experience around the Moon is another benefit to be gained from flying a rehearsal mission before making a lunar landing. More knowledge of the lunar potential or gravitational effect will provide additional refinement of Manned Space Flight Network tracking techniques, and broad landmark tracking will bolster this knowledge.

Analysis of last December's Apollo 8 lunar orbit mission tracking has aided refinements of tracking and navigation techniques and Apollo 10 should reduce error margins still further.

Apollo 10 crewmen are Commander Thomas P. Stafford, Command Module Pilot John W. Young and Lunar Module Pilot Eugene A. Cernan. The mission will be the third space flight for Stafford (Gemini 6 and 9) and Young (Gemini 3 and 10). and the second for Cernan (Gemini 9). The three were recycled from the Apollo 7 backup crew. The Apollo 10 backup crew is Commander L. Gordon Cooper, Command Module Pilot Don F. Eisele and Lunar Module Pilot Edgar D. Mitchell.

Stafford is an Air Force Colonel; Young and Cernan are Navy Commanders.

If necessary, the backup crew can be substituted for the prime crew up to about two weeks prior to an Apollo launch. During this period, the flight hardware and software, ground hardware and software, flight crew and ground crews work as an integrated team to perform ground simulations and other tests of the upcoming mission. It is necessary that the flight crew that will conduct the mission take part in these activities, which are not repeated for the benefit of the backup crew. To do so would add an additional costly two-week period to the prelaunch schedule, which for a lunar mission, would require rescheduling for the next lunar window.

The Apollo 10 rendezvous will be the fifth space rendezvous in which Stafford has taken part. Gemini 7/6 (the world's first rendezvous), and three types of rendezvous with the augmented target docking adapter in Gemini 9.

The Apollo 10 mission time line can be described as a combination of Apollo 8 and Apollo 9 in that it will be a lunar orbit mission with a CSM-LM rendezvous. Apollo 8 was a lunar orbit mission with the command/service module only, while Apollo 9 was an Earth orbital mission with the complete Apollo spacecraft and included a LM-active rendezvous with the CSM.

Apollo 10, after liftoff from launch Complex 39B, will begin the three-day voyage to the Moon about two and a half hours after the spacecraft is inserted into a 100-nautical mile circular Earth parking orbit. The Saturn V launch vehicle third stage will restart to inject Apollo 10 into a translunar trajectory as the vehicle passes over Australia mid-way through the second revolution of the Earth.

The "go" for translunar injection will follow a complete checkout of the spacecraft's readiness to be committed for injection. About an hour after translunar injection (TLI), the command/service module will separate from the Saturn third stages, turn around and dock with the lunar module nested in the spacecraft LM adapter. Spring-loaded lunar module hold-downs will be released to eject the docked spacecraft from the adapter.

Later, leftover liquid propellant in the Saturn third stage will be vented through the engine bell to place the stage into a "slingshot" trajectory to miss the Moon and go into solar orbit,

During the translunar coast, Apollo 10 will be in the so-called passive thermal control mode in which the spacecraft rotates slowly about one of its axes to stabilize thermal response to solar heating. Four midcourse correction maneuvers are possible during translunar coast and will be planned in real time to adjust the trajectory.

Apollo 10 will first be inserted into a 60-by-170 nautical mile elliptical lunar orbit, which two revolutions later will be circularized to 60 nautical miles. Both lunar orbit insertion burns (LOI) will be made when Apollo 10 is behind the Moon out of "sight" of Manned Space Flight Network stations.

Stafford and Cernan will man the LM for systems checkout and preparations for an eight-and-a-half hour sequence that duplicates — except for an actual landing — the maneuvers planned for Apollo 11. The LM twice will sweep within 50,000 feet of Apollo Landing Site 2, one of the prime targets for the Apollo 11 landing.

Maximum separation between the LM and the CSM during the rendezvous sequence will be about 350 miles and will provide an extensive checkout of the LM rendezvous radar as well as of the backup VHF ranging device aboard the CSM, flown for the first time on Apollo 10.

When the LM ascent stage has docked with the CSM and the two crewmen have transferred back to the CSM, the LM will be jettisoned for a ground command ascent engine burn to propellant depletion which will place the LM ascent stage into solar orbit.

The crew of Apollo 10 will spend the remainder of the time in lunar orbit conducting lunar navigational tasks and photographing Apollo landing sites that are within camera range of Apollo 10's ground track.

The transearth injection burn will be made behind the Moon after 61.5 hours in lunar orbit. During the 54-hour transearth coast, Apollo 10 again will control solar heat loads by using the passive thermal control "barbecue" technique. Three transearth midcourse corrections are possible and will be planned in real time to adjust the Earth entry corridor.

Apollo 10 will enter the Earth's atmosphere (400,000 feet) at 191 hours 51 minutes after launch at 36,310 feet-per-second, Command Module touchdown will be 1,285 nautical miles downrange from entry at 15 degrees 7 minutes South latitude by 165 degrees West longitude at an elapsed time of 192 hours 5 minutes. The touchdown point is about 345 nautical miles east of Pago Pago Tutuila, in American Samoa.

(END OF GENERAL RELEASE; BACKGROUND INFORMATION FOLLOWS)

APOLLO 10 — Launch and Trans Lunar Injection

Astronauts Board Apollo

Saturn Staging

Trans Lunar Injection

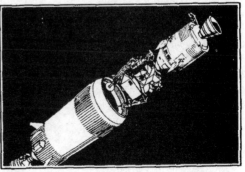

Apollo Saturn Separation

.POLLO 10 — Trans Lunar Flight

Apollo Midcourse Maneuver

Astronauts at Command Module Stations

.POLLO 10 — Trans Lunar Flight

Final Course Adjustment

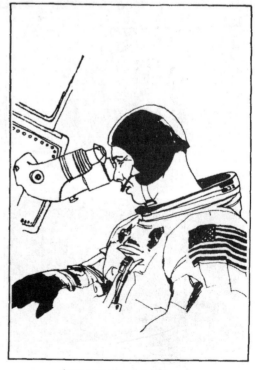

Navigational Check

APOLLO 10 — Lunar Orbital Flight

Lunar Orbit Insertion

Television Broadcast

Lunar Landmark Tracking

Transfer to Lunar Module

APOLLO 10 — Lunar Descent and Rendezvous

Descent Orbit Insertion

Lunar Module Staging

Apollo Docking

LM
Ascent Engine Firing to Depletion

Lunar Landmark Tracking

APOLLO 10 — Trans Earth Injection and Flight

Trans Earth Injection

Apollo Midcourse Maneuver

Navigational Check

Final Reentry Preparations

APOLLO 10 — Earth Reentry and Recovery

Command-Service Module
Separation

Command Module Reentry

Splashdown

Recovery

MISSION OBJECTIVES

Although Apollo 10 will pass no closer than eight nautical miles from the lunar surface, all other aspects of the mission will be similar to the first lunar landing mission, Apollo 11, now scheduled for July.

The trajectory, time line and maneuvers follow the lunar landing profile. After rendezvous is completed, the Apollo 10 time line will deviate from Apollo 11 in that Apollo 10 will spend an extra day in lunar orbit.

Additional LM operation in either Earth orbit or lunar orbit will provide additional experience and confidence with the LM systems, including various control modes of the LM primary/abort guidance systems, as well as further assessment of crew time lines.

The mission will also test the Apollo rendezvous radar at maximum range (approximately 350 miles vs. 100 miles during Apollo 9). Apollo 10 will mark the first space flight test of the LM steerable S-band antenna and of the LM landing radar. The LM landing radar has undergone numerous tests in Earth environment, but this mission will provide an opportunity to check the lunar surface reflectivity characteristics with the landing radar.

Some 800 seconds of landing radar altitude-measuring data will be gathered as the LM makes two sweeps eight nautical miles above Apollo Landing Site 2.

This mission will also provide the first opportunity to check the very high frequency (VHF) ranging device aboard the CSM which serves as a backup to the LM rendezvous radar.

The Apollo 10 mission profile provides fuel and other consumable reserves in the LM that are greater than those planned for the first LM to land on the Moon. The lunar landing mission is the "design mission" for the Apollo spacecraft, and such a mission has smaller although adequate margins of reserve consumables.

From liftoff through descent orbit insertion, Apollo 10 follows closely the trajectory and time line that will be flown in the landing mission. Following the eight-mile pericynthion, the profile closely simulates the conditions of lunar orbit rendezvous after a landing.

The May 18 launch date will produce lighting conditions on Apollo Site 2 similar to those that will be present for the landing mission. At the low inclination to be flown on Apollos 10 and 11 — about 1.2 degrees relative to the lunar equator— Apollo landing site 3 can be photographed and optically tracked by the crew of Apollo 10 in addition to the prime Site 2.

Site 1 was photographed by Apollo 8 in last December's lunar orbit mission and, together with the two sites to be covered in Apollo 10, photographic, tracking and site altitude data on three sites will be in hand.

Among the Apollo 10 objectives is the gathering of additional Manned Space Flight Network (MSFN) tracking data on vehicles in lunar orbit. While MSFN experience in tracking Apollo 8 will benefit Apollo 10 there are still some uncertainties. For example, there is still some lack of knowledge as to what the exact lunar potential or gravity field is and how it affects an orbiting spacecraft.

In tracking Apollo 8, downtrack, or orbital timing errors projected ahead two revolutions were 30,000 feet, and orbital radius measurements relative to the center of the Moon were off 5,500 feet. MSFN tracking can produce accurate position and velocity information in real time while a spacecraft is "in view" from the Earth and not occulted by the Moon, but landing and rendezvous operations will require accurate predictions of position and velocity several revolutions in advance of the event.

The lunar potential apparently affects an orbiting spacecraft differently depending upon orbital inclination and altitude. Apollo 10 will be flown on the same inclination to the lunar equator as the landing mission and will provide information for refining prediction techniques.

Apollo 8 postflight analysis has produced modifications to tracking and position prediction techniques which should reduce downtrack errors to 3,000 feet and altitude errors to 1,400 feet. Apollo 10 will allow mission planners to perfect techniques developed as a result of Apollo 8 tracking analysis.

Other space navigation benefits from Apollo 10 will be gained from combining onboard spacecraft lunar landmark tracking data with MSFN tracking and from evaluating present lunar landing site maps at close visual and camera ranges. Additionally, LM descent and ascent engine burns will be monitored by MSFN stations for developing useful techniques for tracking powered flight in future missions.

APOLLO 10 COUNTDOWN

The clock for the Apollo 10 countdown will start at T-28 hours, with a six hour built-in-hold planned at T-9 hours, prior to launch vehicle propellant loading. The countdown is preceded by a pre-count operation that begins some 4 days before launch. During this period the tasks include mechanical buildup of both the command service module and LM fuel cell activation as well as servicing and loading of the super critical helium aboard the LM descent stage. A 5½ hold is scheduled between the end of the pre-count and start of the final countdown. Following are some of the highlights of the final count:

T-28 hrs.	Official countdown starts
T-27 hrs. 30 mins.	Install launch vehicle flight batteries (to 23 hrs. 30 mins.) LM stowage and cabin closeout (to 15 hrs.)
T-21 hrs.	Top off LM super critical helium (to 19 hrs.)
T-16 hrs.	Launch vehicle range safety checks (to 15 hrs.)
T-11 hrs. 30 mins.	Install launch vehicle destruct devices (to 10 hrs. 45 mins.) Command/service module pre-ingress operations
T-10 hrs.	Start mobile service structure move to park site
T-9 hrs.	Start six hour built-in-hold
T-8 hrs. counting	Clear blast area for propellant loading
T-8 hrs. 30 mins.	Astronaut backup crew to spacecraft for prelaunch checks
T-8 hrs. 15 mins.	Launch Vehicle propellant loading, three stages (liquid oxygen in first stage) liquid oxygen and liquid hydrogen in second, third stages. Continues through T-3 hrs. 38 mins.
T-5 hrs.	Flight crew alerted
T-4 hrs. 45 mins.	Medical examination
T-4 hrs. 15 mins.	Breakfast
T-3 hrs. 45 mins,	Don space suits
T-3 hrs. 30 mins.	Depart Manned Spacecraft Operations Building for LC-39 via crew transfer van
T-3 hrs. 14 mins.	Arrive at LC-39
T-3 hrs. 10 mins.	Enter Elevator to spacecraft level
T-2 hrs. 40 mins.	Start flight crew ingress
T-1 hr. 55 mins.	Mission Control Center-Houston/ spacecraft command checks
T-1 hr. 50 mins.	Abort advisory system checks
T-1 hr. 46 mins.	Space vehicle Emergency Detection System (EDS) test
T-43 mins.	Retrack Apollo access arm to standby position (12 degrees)
T-42 mins.	Arm launch escape system
T-40 mins.	Final launch vehicle range safety checks (to 35 mins.)
T-30 mins.	Launch vehicle power transfer test LM switch over to internal power
T-20 mins. to T-10 mins.	Shutdown LM operational instrumentation
T-15 mins.	Spacecraft to internal power
T-6 mins.	Space vehicle final status checks
T-5 mins. 30 sec.	Arm destruct system
T-5 mins.	Apollo access arm fully retracted
T-3 mins. 10 sec.	Initiate firing command (automatic sequencer)
T-50 sec.	Launch vehicle transfer to internal power
T-8.9 sec*	Ignition sequence start
T - 2 sec*	All engines running
T - 0	Liftoff

Note: Some changes in the above countdown are possible, as a result of experience gained in the Countdown Demonstration Test (CDDT) which occurs about 10 days before launch.

MISSION TRAJECTORY AND MANEUVER DESCRIPTION

(Note, information presented herein is based upon a May 16 launch and is subject to change prior to the mission or in real time during the mission to meet changing conditions.)

Launch

Apollo 10 will be launched from Kennedy Space Center launch Complex 39B on a launch azimuth that can vary from 72 degrees to 108 degrees, depending upon the time of day of launch. The azimuth changes with time of day to permit a fuel-optimum injection from Earth parking orbit into a free-return circumlunar trajectory. Other factors influencing the launch windows are a daylight launch and proper Sun angles on lunar landing sites.

The planned Apollo 10 launch date of May 18 will call for liftoff at 12:49 p.m. EDT on a launch azimuth of 72 degrees. Insertion into a 100-nautical-mile circular Earth parking orbit will occur at 11 minutes 53 seconds ground elapsed from launch (GET), and the resultant orbit will be inclined 32.5 degrees to the Earth's equator.

FLIGHT PROFILE

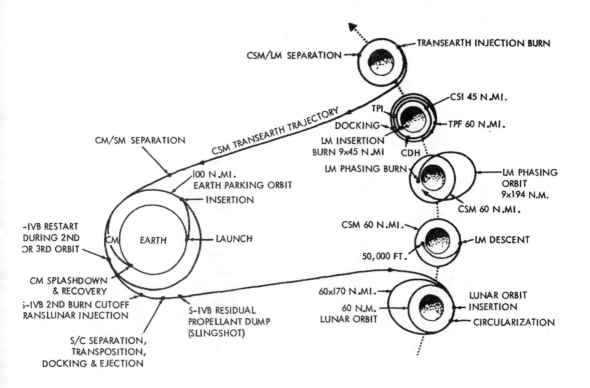

SPACE VEHICLE LAUNCH EVENTS/ WEIGHTS

Hrs.	Time Min.	Sec.	Event	Altitude Naut. Mi.	Velocity Knots	Weight Pounds
00	00	(-)08.9	Ignition	0.00	0	6,499,016
00	00	00	First Motion	0.033	*0	6,412,918
00	00	12	Tilt Initiation	0.12	*3	
00	01	21	Maximum Dynamic Pressure	7	1554	
00	02	15	Center Engine Cutoff	24	3888	2,434,985
00	02	40	Outboard Engines Cutoff	35	5324	1,842,997
00	02	41	S-IC/ S-II Separation	36	5343	1,465,702
00	02	42	S-II Ignition	37	5335	1,465,123
00	03	11	S-II Aft Interstage Jettison	49	5581	
00	03	16	LES Jettison	51	5642	
00	03	21	Initiate IGM	53	5701	-
00	07	39	S-II Center Engine Cutoff	97	10977	644,128
00	09	14	S-II Outboard Engines Cutoff	102	13427	471,494
00	09	15	S-II/ S-IVB Separation	102	13434	364,429
00	09	18	S-IVB Ignition	102	13434	364,343
00	11	43	S-IVB First Cutoff	103	15135	295,153
00	11	53	Parking Orbit Insertion	103	15139	295,008

*First two velocities are space fixed, Others are inertial velocities. Vehicle on launch pad has inertial velocity of 408.5 meters per second (793.7 knots).

The above figures are based on a launch azimuth of 72 degrees. Figures will vary slightly for other azimuths.

Apollo 10 Mission Events

Event	Ground Elapsed Time. Hr:min:sec	Date & Time (EDT)	Velocity Change ft per sec	Purpose and (Resultant orbit)
Insertion	00:11:53	5/18 1:01 pm	25,593	Insertion into 100 nm Circular EPO.
Translunar injection	02:33:26	5/18 3:23 pm	10,058	Injection into free-return translunar trajectory with 60 nm pericynthion.
CSM separation, docking	03:10:00	5/18 3:59 pm	—	Hard-mating or CSM and LM.
Ejection from SLA	04:09:00	5/18 4:58 pm	1	Separates CSM-LM from S-IVB/ SLA.
SPS evasive maneuver	04:29:00	5/18 5:18 pm	19.7	Provides separation prior to S-IVB Propellant dump and "slingshot" Maneuver.
Midcourse correction No. 1	TLI +9 hrs	5/19 12:22 am	55*	*These midcourse Corrections have a nominal velocity change of 0 fps, but will be calculated in real time to correct TLI dispersions. MCC-3 will have a plane change component to achieve desired lunar orbit inclination.
Midcourse correction No. 2	TLI +24 hrs	5/19 3:23 pm	0*	
Midcourse correction No. 3	LOI -22 hrs	5/20 6:35 pm	0*	
Midcourse correction No. 4	LOI -5 hrs	5/21 11:35 am	0*	
Lunar Orbit Insertion No. 1	75:45:43	5/21 4:35 pm	-2,974	Inserts Apollo 10 into 60 x 170 nm elliptical lunar orbit.
Lunar Orbit Insertion No. 2	80:10:45	5/21 9:00 pm	-138.5	Circularizes lunar parking orbit to 60 nm.
CSM- LM undocking; separation (SM RCS)	98:10:00 / 98:35:26	5/22 2:59 pm / 5/22 3:24 pm	2.5	Establishes equiperiod orbit for 2 nm separation (mini-football).
Descent orbit Insertion (DPS)	99:33:59	5/22 4:23 pm	-71	Lower LM Pericynthion to eight nm (8x60) .
DPS phasing burn	100:46:21	5/22 5:35 pm	195	Raises LM apocynthion to194 nm, allows CSM to pass and overtake LM (8x194)

Event	Ground Elapsed Time. Hr:min:sec	Date & Time (EDT)	Velocity Change ft per sec	Purpose and (Resultant orbit)
APS Insertion burn	102:43:18	5/22 7:32 pm	-207	Simulates LM ascent into lunar orbit after landing (8x43.6)
LM RCS concentric sequence initiate (CSI) burn	103:33:46	5/22 8:22 pm	50.5	Raises LM pericynthion to 46.2 nm, adjusts orbital shape for rendezvous sequence (42.9 x 46.2)
LM RCS constant delta height (CDH) burn	104:31:42	5/22 9:20 pm	3.4	Radially downward burn adjusts LM to constant 15 nm below CSM
LM RCS terminal phase initiate (TPI) burn	105:09:00	5/22 9:58 pm	24.6	LM thrusts along line-of-sight toward CSM, midcourse and braking maneuvers as necessary.
Rendezvous (TPF)	105:54:00	5/22 10:43 pm		Completes rendezvous sequence. Fly formation at 100 ft.
Docking	106:20:00	5/22 11:09 pm		Transfer back to CSM (about 107 GET).
APS burn to depletion	108:38:57	5/23 1:28 am	3,837	Posigrade APS depletion burn near LM pericynthion injects LM ascent stage into heliocentric orbit.
Transearth Injection (TEI) SPS burn.	137:20:22	5/24 2:09 am	3,622.5	Injects CSM into 54½ hour transearth trajectory.
Midcourse correction No.5 Midcourse correction No. 6 Midcourse correction No. 7	TEI +15 hrs. Entry - 15 hrs. Entry - 3 hrs	5/24 5:09 pm 5/25 5:39 pm 5/26 5:39 am	— — —	Transearth Midcourse Corrections will be computed in real time for entry corridor control and for adjusting landing point to avoid recovery area foul weather.
CM/SM separation	191:35	5/26 8:24 am	—	Reentry condition.
Entry interface (400,000 feet)	191:50:32	5/26 8:39 am	—	Command module enters Earth's sensible atmosphere at 36,310 fps.
Touchdown	192:04:47	5/26 8:54 am	—	Landing 1,285 nm downrange from entry 15 degrees seven minutes South latitude x 165 degrees West longitude.

The crew for the first time will have a backup to launch vehicle guidance during powered flight. If the Saturn instrument unit inertial platform fails, the crew can switch guidance to the command module computer for first-stage powered flight automatic control. Second and third stage backup guidance is through manual takeover in which command module hand controller inputs are fed through the command module computer to the Saturn instrument unit.

Earth Parking Orbit (EPO)

Apollo 10 will remain in Earth parking orbit for one-and-one-half revolutions after insertion and will hold a local horizontal attitude during the entire period. The crew will perform spacecraft systems checks in preparation for the translunar injection (TLI) burn. The final "go" for the TLI burn will be given to the crew through the Carnarvon, Australia, Manned Space Flight Network station.

Translunar Injection (TLI)

Midway through the second revolution in Earth parking orbit, the S-IVB third-stage engine will reignite at two hours 33 minutes 26 seconds Ground Elapsed Time (GET) over Australia to inject Apollo 10 toward the Moon. The velocity will increase from 25,593 feet-per-second (fps) to 35,651 fps at TLI cutoff — a velocity increase of 10,058 fps. The TLI burn will place the spacecraft on a free-return circumlunar trajectory from which midcourse corrections could be made with the SM reaction control system thruster. Splashdown for a free-return trajectory would be at 6:37 p.m. EDT May 24 at 24.9 degrees South latitude by 84.3 degrees East longitude after a flight time of 149 hours and 49 minutes.

Transposition, Docking and Ejection (TD&E)

At about three hours after liftoff and 25 minutes after the TLI burn, the Apollo 10 crew will separate the command/ service module from the spacecraft lunar module adapter (SLA), thrust out away from the S-IVB, turn around and move back in for docking with the lunar module. Docking should take place at about three hours and ten minutes GET, and after the crew confirms all docking latches solidly engaged, they will connect the CSM-to-LM umbilicals and pressurize the LM with the command module surge tank. At about 4:09 GET, docked spacecraft will be ejected from the spacecraft LM adapter by spring devices at the four LM landing gear "knee" attach points. The ejection springs will impart about one fps velocity to the spacecraft. A 19.7 fps service propulsion system (SPS) evasive maneuver in plane at 4:29 GET will separate the spacecraft to a safe distance for the S-IVB "slingshot" maneuver in which residual liquid propellants will be dumped through the J-2 engine bell to propel the stage into a trajectory passing behind the Moon's trailing edge and on into solar orbit.

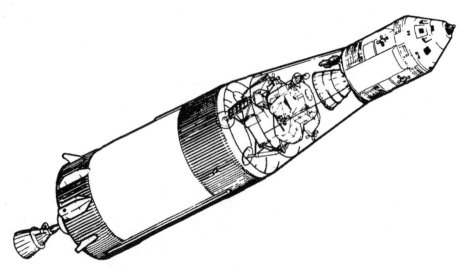

SPACE VEHICLE EARTH PARKING ORBIT CONFIGURATION

(SATURN V THIRD STAGE AND INSTRUMENT UNIT, APOLLO SPACECRAFT)

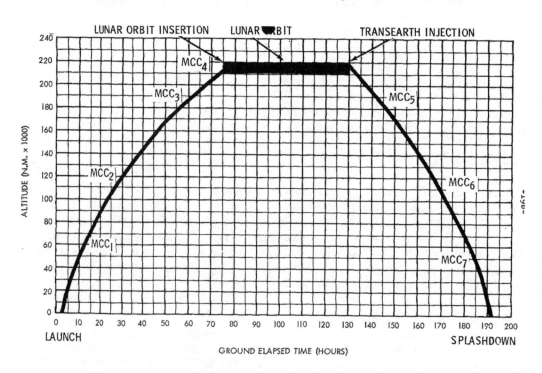

SPACECRAFT ALTITUDE VS. TIME

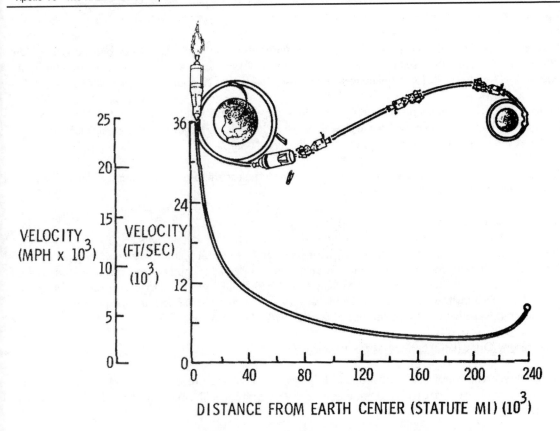

VELOCITY (MPH x 10³)

VELOCITY (FT/SEC) (10³)

DISTANCE FROM EARTH CENTER (STATUTE MI) (10³)

TRANSLUNAR VELOCITY PROFILE

POST TLI TIMELINE

TLI + 20 SEC
LOCAL
HORIZONTAL
ORBIT RATE

TLI + 15 MIN
MANEUVER TO
SEPARATION
ATTITUDE

TLI + 25 MIN SC
INITIAL SEPARATION
(1 FPS)

TLI + 27 MIN
NULL SEPARATION
RATE AND PITCH
TO DOCKING
ATTITUDE

TLI + 90 MIN
LM WITHDRAWAL

TLI + 110 MIN

SPS EVASIVE
MANEUVER
~20 FPS

Translunar Coast

Up to four midcourse correction burns are planned during the translunar coast phase, depending upon the accuracy of the trajectory resulting from the TLI maneuver. If required, the midcourse correction burns are planned at TLI +9 hours, TLI +24 hours, lunar orbit insertion (LOI) -22 hours and LOI -5 hours.

During coast periods between midcourse corrections, the spacecraft will be in the passive thermal control (PTC) or "barbecue" mode in which the spacecraft will rotate slowly about one axis to stabilize spacecraft thermal response space to the continuous solar exposure.

Midcourse corrections 1 and 2 will not normally be made unless the predicted Mission Control Center 3 velocity change is greater than 25 feet-per-second.

Lunar Orbit Insertion (LOI)

The first of two lunar orbit insertion burns will be made at 75:45:43 GET at an altitude of 89 nm above the Moon. LOI-1 will have a nominal retrograde velocity change of 2,974 fps and will insert Apollo 10 into a 60 x 170-nm elliptical lunar orbit. LOI-2 two orbits later at 80:10:45 GET will circularize the orbit to 60 nm. The burn will be 138.5 fps retrograde. Both LOI maneuvers will be with the SPS engine near pericynthion when the spacecraft is behind the Moon and out of contact with MSFN stations.

Lunar Parking Orbit (LPO) and LM-Active Rendezvous

Apollo 10 will remain in lunar orbit about 61.5 hours, and in addition to the LM descent to eight nautical miles above the lunar surface and subsequent rendezvous with the CSM, extensive lunar landmark tracking tasks will be performed by the crew.

Following a rest period after the lunar orbit circularization, the LM will be manned by the command and lunar module pilot and preparations begun for undocking at 98:10 GET. Some 25 minutes of station keeping and CSM inspection of the LM will be followed by a 2.5 fps radially downward SM RCS maneuver, placing the LM and CSM in equiperiod orbits with a maximum separation of two miles (minifootball). At the midpoint of the minifootball, the LM descent propulsion system (DPS) will be fired retrograde 71 fps at 99:34 GET for the descent orbit insertion (DOI) to lower LM pericynthion to eight miles. The DPS engine will be fired at 10 per cent throttle setting for 15 seconds and at 40 per cent for 13 seconds.

LUNAR ORBIT INSERTION

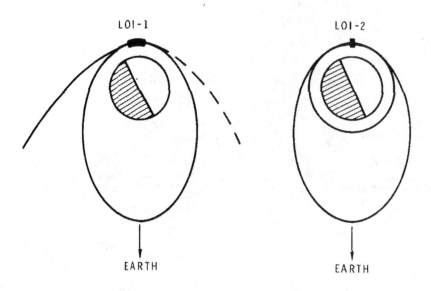

LOI-1   LOI-2

EARTH   EARTH

LUNAR ORBIT ACTIVITIES

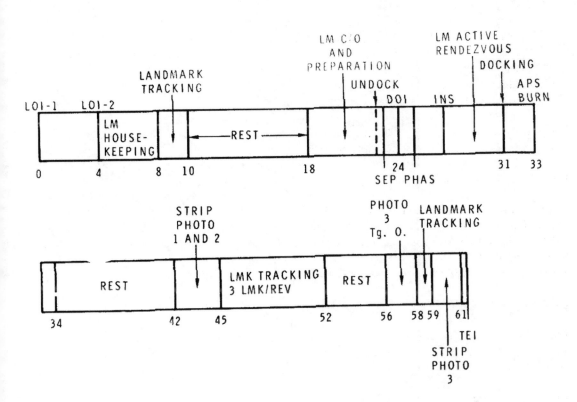

COMPARISON OF F AND G LM OPERATIONS PHASE

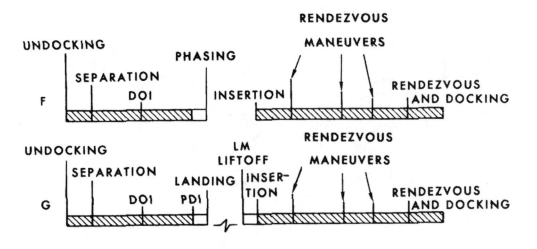

CSM/LM TYPICAL LANDMARK TRACKING PROFILE

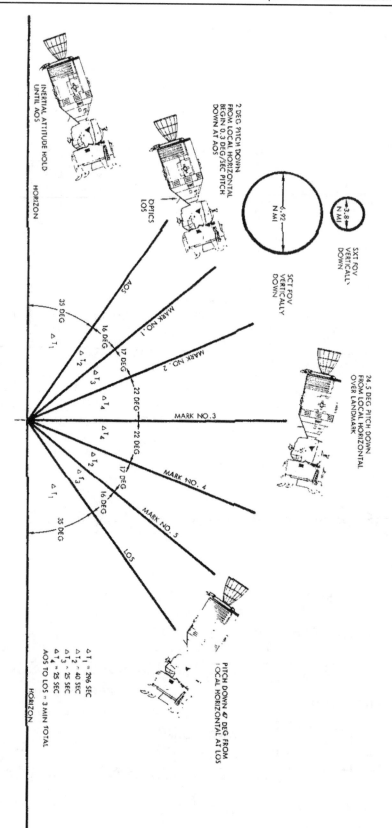

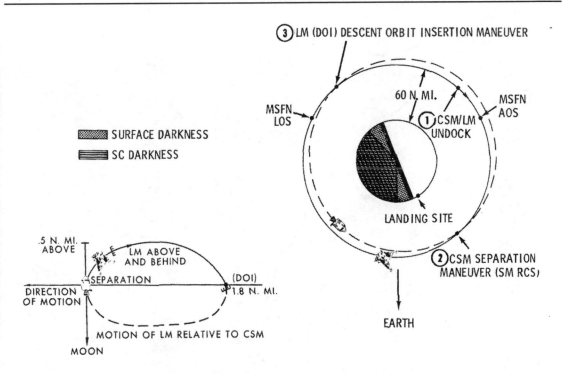

CSM/LM SEPARATION MANEUVER

LUNAR MODULE DESCENT ORBIT INSERTION

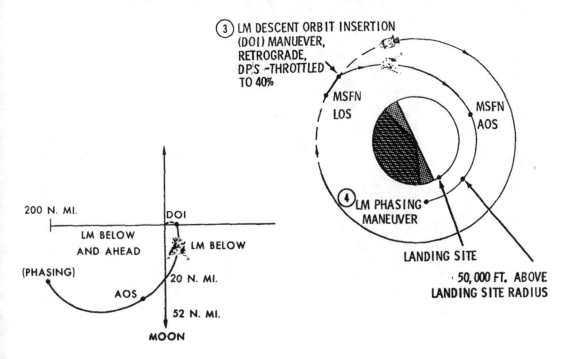

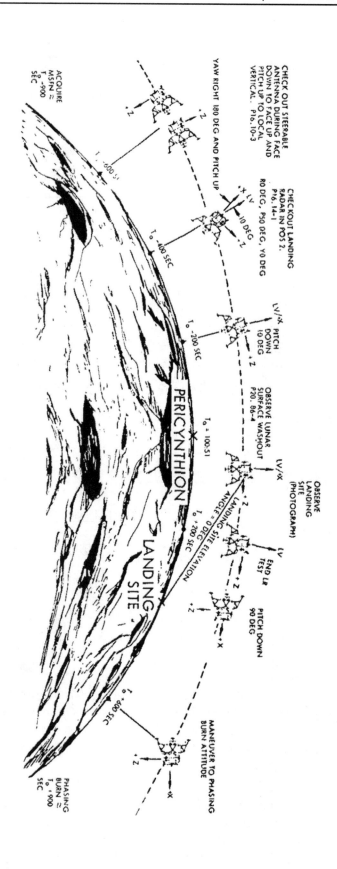

NEAR LUNAR SURFACE ACTIVITY

ORBIT RATE (0.05 DEG/SEC PITCH DOWN) FROM -400 to 200 FROM PERICYTHION

APOLLO 10 RENDEZVOUS SEQUENCE

MANEUVER	G.E.T.	deltaV, FPS	ENGINE
SEPARATION	98:35:23	2.5	SM RCS
DOI	99:33:59	71.0	DPS
PHASING	100:46:21	195.0	DPS
INSERTION	102:43:18	207.0	APS
CSI	103:33:46	50.5	LM RCS
CDH	104:31:42	3.4	LM RCS
TPI	105:09:00	24.8	LM RCS
BRAKING	~105:55:00	~60.0	LM RCS
DOCKING	~106:20:00	~5.0	SM RCS

As the LM passes over Apollo landing Site 2, the LM landing radar will be tested in the altitude mode but not in descent rate. About 10 minutes after the pass over Site 2, the 195 fps DPS phasing burn at 100:46 GET will boost the LM into an 8 x 194 nm orbit to allow the CSM to overtake and pass the LM. The phasing burn is posigrade and the DPS engine is fired at 10 per cent throttle for 26 seconds and full throttle for 17 seconds. The phasing burn places the LM in a "dwell" orbit which allows the CSM to overtake and pass the LM so that at the second LM passes over Site 2, the LM will trail the CSM by 27 nm and will be in a proper position for the insertion maneuver simulating ascent from the lunar surface after a landing mission.

Prior to the 207-fps LM ascent engine retrograde insertion burn, the LM descent stage will be jettisoned and an evasive maneuver performed by the ascent stage to prevent recontact. The insertion burn will be made at 102:43 GET and will lower 148 apocynthion to 44.9 nm so that the LM is 14.7 nm below and 148 nm behind the CSM at the time of the concentric sequence initiate (CSI) burn.

Following LM radar tracking of the CSM and onboard computation of the CSI maneuver, a 50.5 fps LM RCS posigrade burn will be made at a nominal time of 103:33 GET at apocynthion and will result in a 44.9 x 44.3 nm LM orbit. The LM RCS will draw from the LM ascent propulsion system (APS) propellant tanks through the interconnect valves.

A 3.4 fps radially downward LM RCS constant delta height (CDH) maneuver at 104:31 GET will place the LM on a coelliptic orbit 15 nm below that of the CSM and will set up conditions for the terminal phase initiate (TPI) burn 38 minutes later.

The TPI maneuver will be made when the CSM is at a 26.6 degree elevation angle above the LM's local horizontal following continuing radar tracking of the CSM and onboard computations for the maneuver. Nominally, the TPI burn will be a 24.6-fps LM RCS burn along the line of sight toward the CSM at 105:09 GET. Midcourse correction and braking maneuvers will place the LM and CSM in a rendezvous and station-keeping position, and docking should take place at 106:20 GET to complete a eight-and-a-half hour sequence of undocked activities.

After the commander and lunar module pilot have transferred into the CSM, the LM will be jettisoned and the CSM will maneuver 2 fps radially upward to move above and behind the LM at the time of the LM ascent propulsion system burn to propellant depletion at 108:39 GET.

LUNAR MODULE PHASING MANEUVER

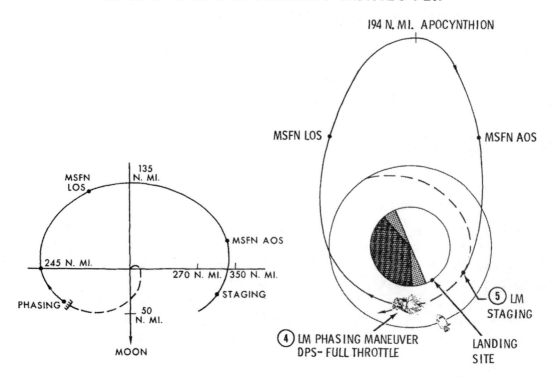

LUNAR MODULE INSERTION MANEUVER

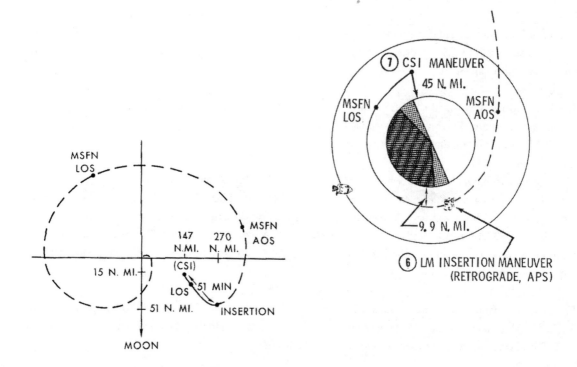

LUNAR MODULE
CONCENTRIC SEQUENCE INITIATION MANEUVER

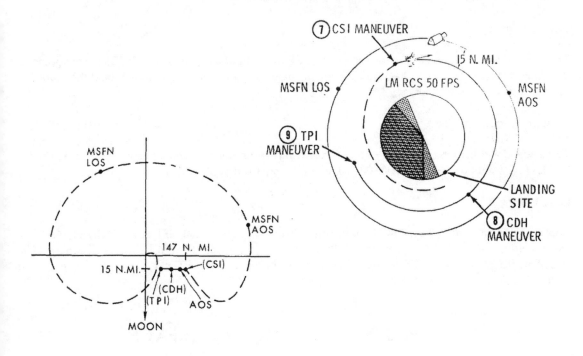

LUNAR MODULE CONSTANT
DIFFERENTIAL HEIGHT AND TERMINAL PHASE MANEUVERS

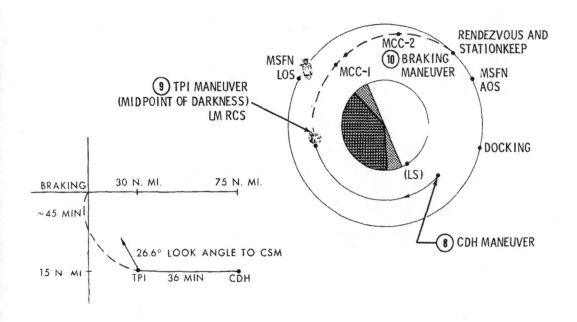

The burn will be ground-commanded. An estimated 3,837-fps posigrade velocity will be imparted by the APS depletion burn near LM pericynthion to place the LM ascent stage in a heliocentric orbit.

An additional 29 hours will be spent in lunar orbit before transearth injection while the crew conducts lunar landmark tracking tasks and makes photographs of Apollo landing sites.

Transearth Injection (TEI)

The 54-hour return trip to Earth begins at 137:20 GET when the SPS engine is fired 3622.5 fps posigrade for the TEI burn. Like LOI-1 and LOI-2, the TEI burn will be made when the spacecraft is behind the Moon and out of touch with MSFN stations.

Transearth Coast

Three corridor-control transearth midcourse correction burns will be made if needed: MCC-5 at TEI +15 hours, MCC-6 at entry interface (EI=400,000 feet) -15 hours and at EI -3 hours.

Entry, Landing

Apollo 10 will encounter the Earth's atmosphere (400,000 feet) at 191:50 GET at a velocity of 36,310 fps and will land some 1,285 nm downrange from the entry-interface point using the spacecraft's lifting characteristics to reach the landing point. Touchdown will be at 192:05 GET at 15 degrees 7 minutes South latitude by 165 degrees West longitude.

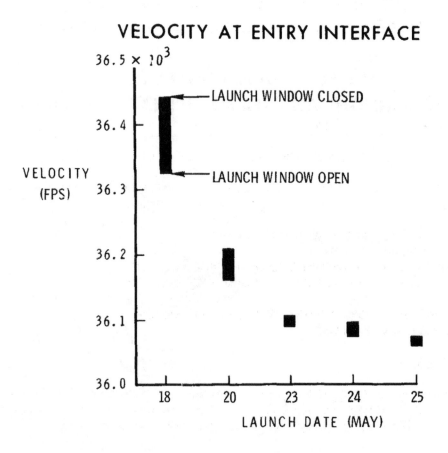

GEODETIC ALTITUDE VERSUS RANGE TO GO

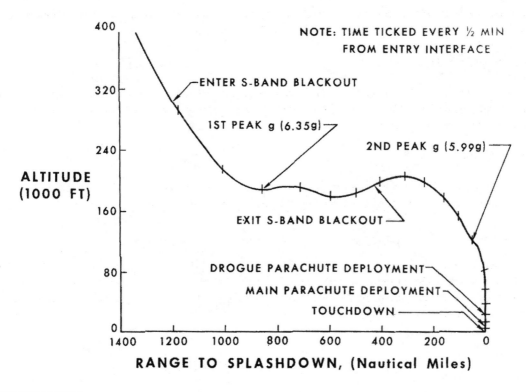

RANGE TO SPLASHDOWN, (Nautical Miles)

EARTH ENTRY

ENTRY RANGE CAPABILITY - 1200 TO 2500 N. MI.

NOMINAL ENTRY RANGE - 1285 N. Mi.

- SHORT RANGE SELECTED FOR NOMINAL MISSION BECAUSE:

- RANGE FROM ENTRY TO LANDING CAN BE SAME FOR PRIMARY AND BACKUP CONTROL MODES

- PRIMARY MODE EASIER TO MONITOR WITH SHORT RANGE

- WEATHER AVOIDANCE, WITHIN ONE DAY PRIOR TO ENTRY, IS ACHIEVED USING ENTRY RANGING CAPABILITY TO 2500 N. MI.

- UP-TO ONE DAY PRIOR TO ENTRY USE PROPULSION SYSTEM TO CHANGE LANDING POINT

RECOVERY OPERATIONS

The primary recovery line for Apollo 10 is in the mid-Pacific along the 175th West meridian of longitude above 15 degrees North latitude, and jogging to 165 degrees West longitude below the Equator. The helicopter carrier USS Princeton, Apollo 10 prime recovery vessel, will be stationed near the end-of-mission aiming point.

Splashdown for a full-duration lunar orbit mission launched on time May 18 will be at 5 degrees 8 minutes South by 165 degrees West at a ground elapsed time of 192 hours 5 minutes.

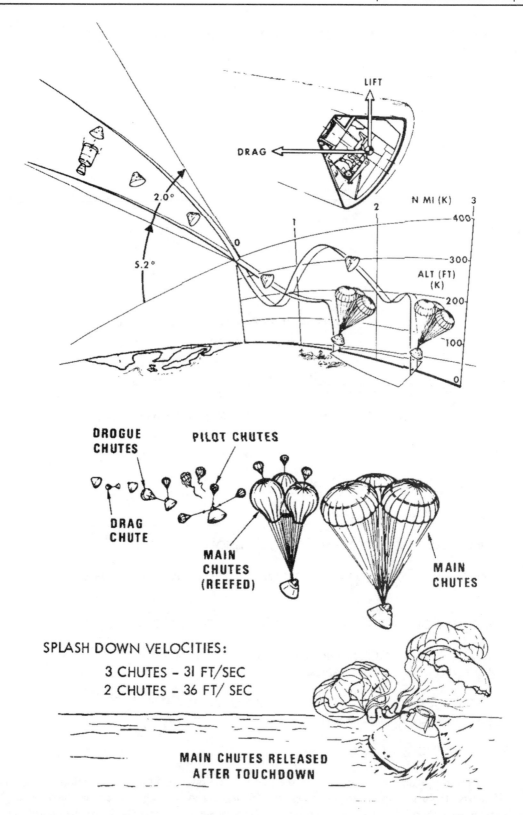

LIFT

DRAG

N MI (K)

2.0°

5.2°

ALT (FT)
(K)

DROGUE
CHUTES

PILOT CHUTES

DRAG
CHUTE

MAIN
CHUTES
(REEFED)

MAIN
CHUTES

SPLASH DOWN VELOCITIES:

3 CHUTES – 31 FT/SEC
2 CHUTES – 36 FT/ SEC

MAIN CHUTES RELEASED
AFTER TOUCHDOWN

EARTH RE-ENTRY AND LANDING

The latitude of splashdown depends upon the time of the transearth injection burn and the declination of the Moon at the time of the burn. A spacecraft returning from a lunar flight will enter Earth's atmosphere and splash down at a point on Earth directly opposite the Moon.

This point, called the antipode, is a projection of a line from the center of the Moon through the center of the Earth to the surface opposite the Moon. The mid-Pacific recovery line rotates through the antipode once each 24 hours, and the trans Earth injection burn will be targeted for splashdown along the primary recovery line.

Other planned recovery lines for a deep-space mission are the East Pacific line extending roughly parallel to the coastlines of North and South America; the Atlantic Ocean line running along the 30th West meridian in the northern hemisphere and along the 25th West meridian in the southern hemisphere; the Indian Ocean line along the 65th East meridian; and the West Pacific line along the 150th East meridian in the northern hemisphere and jogging to the 170th East meridian in the southern hemisphere.

Secondary landing areas for a possible Earth orbital alternate mission have been established in two zones — one in the Pacific and one in the Atlantic.

Launch abort landing areas extend downrange 3,400 nautical miles from Kennedy Space Center, fanwise 50 nautical miles above and below the limits of the variable launch azimuth (72 degrees 107 degrees). Ships on station in the launch abort area will be the destroyer USS Rich, the insertion tracking ship USNS Vanguard and the attack transport USS Chilton.

In addition to the primary recovery vessel steaming up and down the mid-Pacific recovery line and surface vessels on the Atlantic Ocean recovery line and in the launch abort area, 14 HC-130 aircraft will be on standby at seven staging bases around the Earth: Guam, Pago Pago, American Samoa; Hawaii, Bermuda; Lajes, Azores; Ascension Island; Mauritius and the Panama Canal Zone.

Apollo 10 recovery operations will be directed from the Recovery Operations Control Room in the Mission Control Center and will be supported by the Atlantic Recovery Control Center, Norfolk., Va., and the Pacific Recovery Control Center, Kunia, Hawaii.

The Apollo 10 crew will be flown from the primary recovery vessel to the Manned Spacecraft Center after recovery. The spacecraft will receive a preliminary examination, safing and power-down aboard the Princeton prior to offloading at Ford Island, Hawaii, where the spacecraft will undergo a more complete deactivation. It is anticipated that the spacecraft will be flown from Ford Island to Long Beach, Calif., within 72 hours, and then trucked to the North American Rockwell Space Division plant in Downey, Calif., for postflight analysis.

APOLLO 10 ALTERNATE MISSIONS

Five alternate mission plans have been prepared for the Apollo 10, each depending upon when in the mission time line it becomes necessary to switch to the alternate. Testing of the lunar module and a LM-active rendezvous in Earth orbit are preferred over a CSM-only flyby mission. When it is impossible to return to a low Earth orbit with rendezvous, a high-ellipse LM test is preferred over a low Earth orbit test.

Where possible, Apollo 10 alternate missions follow the lunar orbit mission time line and have a duration of about 10 days.

Apollo 10 alternate missions are summarized as follows:

Alternate 1:

Early shutdown of S-IVB during TLI with resulting apogee less than 25,000 nautical miles, or failure of S-IVB to insert spacecraft into Earth parking orbit and subsequent SPS contingency orbit insertion (COI), and in both cases no LM extraction possible. Alternate maneuvers would include:

* SPS phasing burn to obtain ground coverage of simulated lunar orbit insertion.

* Simulated LOI burn to a 100 x 400 nm Earth orbit.

* Midcourse corrections to modify orbit to 90 x 240 nm end-of-mission ellipse and to complete SPS lunar mission duty cycle during remainder of ten-day mission.

Alternate 2:

S-IVB fails during TLI burn and resulting apogee is between 25,000 and 40,000 nautical miles; no LM extraction. Maneuver sequence would be:

* SPS phasing burn to obtain ground coverage of simulated lunar orbit insertion.

* Simulated LOI burn to a semi-synchronous Earth orbit.

* SPS phasing maneuver to place a later perigee over or opposite desired recovery zone.

* SPS maneuver to place CSM in semi-synchronous orbit with a 12-hour period.

* Deorbit directly from semi-synchronous orbit into Pacific recovery area (ten-day mission).

Alternate 3:

No TLI burn or TLI apogee less than 4,000 nm but LM successfully extracted.

* Simulated LOI burn to 100 x 400-nm orbit.

* Simulated descent orbit insertion (DOI) maneuver with LM.

* Simulated LM powered descent initiation (PDI) maneuver.

* Two SPS burns to circularize CSM orbit to 150 nm.

* LM-active rendezvous.

* Ground-commanded LM ascent propulsion system (APS) burn to depletion under abort guidance system (AGS) control, similar to APS depletion burn in Apollo 9.

* Additional SPS burns to place CSM in 90 x 240 nm end of mission ellipse and to complete SPS lunar mission duty cycle during remainder of ten-day mission.

Alternate 4:

Early S-IVB TLI cutoff with resulting apogee greater than 4,000 nm but less than 10,000 nm, and capability of SPS and LM descent propulsion system together to return CSM-LM to low Earth orbit without compromising CSM's ability to rescue LM.

* SPS phasing burn to obtain ground coverage of simulated lunar orbit insertion.

* First docked DPS burn out-of-plane simulates descent orbit insertion.

* Second docked DPS burn simulates power descent initiation.

* SPS simulated LOI burn.

* Phasing maneuver to obtain ground coverage of simulated powered descent initiation.

* SPS burns to circularize CSM orbit at 150 nm.

* LM-active rendezvous.

* Ground-commanded LM ascent propulsion system burn to depletion under abort guidance system (AGS) control, similar to APS depletion burn in Apollo 9.

* Additional SPS burns to place CSM in 90 x 240 nm end of mission ellipse and to complete SPS lunar mission duty cycle during remainder of ten-day mission.

Alternate 5:

SPS and DPS jointly cannot place CSM-LM in low Earth orbit without compromising ability of CSM to rescue LM in a rendezvous sequence, and SPS fuel quantity is too low for a CSM-LM circumlunar mission.

* SPS phasing burn to obtain ground coverage of simulated lunar orbit insertion.

* Simulated lunar orbit insertion into semisynchronous orbit.

* SPS phasing burn to obtain ground coverage of simulated power descent initiation.

* First docked DPS burn out of plane simulates descent orbit insertion.

* Second docked DPS burn simulates power descent initiation and is directed out of plane.

* SPS phasing burn to place a later perigee over or opposite desired recovery zone.

* SPS maneuver to place CSM-LM in semi-synchronous orbit with a 12-hour period.

* Ground-commanded LM ascent propulsion system burn to depletion under abort guidance system control; posigrade at apogee.

* Additional midcourse corrections along a lunar mission time line and direct entry from high ellipse.

ABORT MODES

The Apollo 10 mission can be aborted at any time during the launch phase or terminated during later phases after a successful insertion into Earth orbit.

Abort modes can be summarized as follows:

Launch phase

Mode I - launch escape (LES) tower propels command module away from launch vehicle. This mode is in effect from about T-45 minutes when LES is armed until LES jettison at 3:07 GET and command module landing point can range from the Launch Complex 39B area to 520 nm (600 sm, 964 km) downrange.

Mode II - Begins when LES is jettisoned and runs until the SPS can be used to insert the CSM into a safe Earth orbit (9:22 GET) or until landing points threaten the African coast. Mode II requires manual separation, entry orientation and full lift entry with landing between 400 and 3,200 nm (461-3,560 sm, 741-5,931 km) downrange.

Mode III - Begins when full-lift landing point reaches 3,200 nm (3,560 sm, 5,931 km) and extends through Earth orbital insertion. The CSM would separate from the launch vehicle, and if necessary, an SPS retrograde burn would be made, and the command module would be flown half-lift to entry and landing at approximately 3,350 nm (3,852 sm, 6,197 km) downrange.

Mode IV and Apogee Kick - Begins after the point the SPS could be used to insert the CSM into an Earth parking orbit — from about 9:22 GET. The SPS burn into orbit would be made two minutes after separation from the S-IVB and the mission would continue as an Earth orbit alternate. Mode IV is preferred over Mode III. A variation of Mode IV is the apogee kick in which the SPS would be ignited at first apogee to raise perigee for a safe orbit.

Deep Space Aborts

Translunar Injection Phase

Aborts during the translunar injection phase are only a remote possibility, but if an abort became necessary during the TLI maneuver, an SPS retrograde burn could be made to produce spacecraft entry. This mode of abort would be used only in the event of an extreme emergency that affected crew safety. The spacecraft landing point would vary with launch azimuth and length of the TLI burn. Another TLI abort situation would be used if a malfunction cropped up after injection. A retrograde SPS burn at about 90 minutes after TLI shutoff would allow targeting to land on the Atlantic Ocean recovery line.

Translunar Coast phase

Aborts arising during the three-day translunar coast phase would be similar in nature to the 90-minute TLI abort. Aborts from deep space bring into play the Moon's antipode (line projected from Moon's center through Earth's center to opposite face) and the effect of the Earth's rotation upon the geographical location of the antipode. Abort times would be selected for landing when the antipode crosses 165° West longitude. The antipode crosses the mid-Pacific recovery line once each 24 hours, and if a time-critical situation forces an abort earlier than the selected fixed abort times, landings would be targeted for the Atlantic Ocean, East Pacific, West Pacific or Indian Ocean recovery lines in that order of preference. When the spacecraft enters the Moon's sphere of influence, a circumlunar abort becomes faster than an attempt to return directly to Earth.

Lunar Orbit Insertion phase

Early SPS shutdowns during the lunar orbit insertion burn (LOI) are covered by three modes in the Apollo 10 mission. All three modes would result in the CM landing at the Earth latitude of the Moon antipode at the time the abort was performed.

Mode I would be a LM DPS posigrade burn into an Earth-return trajectory about two hours (at next pericynthion) after an LOI shutdown during the first two minutes of the LOI burn.

Mode II, for SPS shutdown between two and three minutes after ignition, would use the LM DPS engine to adjust the orbit to a safe, non-lunar impact trajectory followed by a second DPS posigrade burn at next pericynthion targeted for the mid-Pacific recovery line.

Mode III, from three minutes after LOI ignition until normal cutoff, would allow the spacecraft to coast through one or two lunar orbits before doing a DPS posigrade burn at pericynthion targeted for the mid-Pacific recovery line.

Lunar Orbit Phase

If during lunar parking orbit it became necessary to abort, the transearth injection (TEI) burn would be made early and would target spacecraft landing to the mid-Pacific recovery line.

Transearth Injection phase

Early shutdown of the TEI burn between ignition and two minutes would cause a Mode III abort and a SPS posigrade TEI burn would be made at a later pericynthion. Cutoffs after two minutes TEI burn time would call for a Mode I abort —- restart of SPS as soon as possible for Earth-return trajectory. Both modes produce mid-Pacific recovery line landings near the latitude of the antipode at the time of the TEI burn.

Transearth Coast phase

Adjustments of the landing point are possible during the transearth coast through burns with the SPS or the service module RCS thrusters, but in general, these are covered in the discussion of transearth midcourse corrections. No abort burns will be made later than 24 hours prior to entry to avoid effects upon CM entry velocity and flight path angle.

APOLLO 10 GO/NO-GO DECISION POINTS

Like Apollo 8, Apollo 10 will be flown on a step-by-step commit point or go/no-go basis in which the decisions will be made prior to each major maneuver whether to continue the mission or to switch to one of the possible alternate missions. The go/no-go decisions will be made by the flight control teams in Mission Control Center.

Go/no-go decisions will be made prior to the following, events:

* Launch phase go/no-go at 10 min. GET for orbit insertion

* Translunar injection

* Transposition, docking and LM extraction

* Each translunar midcourse correction burn

* Lunar orbit insertion burns Nos. 1 and 2

* Crew intravehicular transfer to LM

* CSM-LM undocking and separation

* Rendezvous sequence

* LM Ascent Propulsion system burn to depletion

* Transearth injection burn (no-go would delay TEI one or more revolutions to allow maneuver preparations to be completed.)

* Each transearth midcourse correction burn

ONBOARD TELEVISION

On Apollo 10, onboard video will originate from the CM; there will, be no TV camera in the LM. Plans call for both black and white and color TV to be carried.

The black and white camera is a 4.5 pound RCA camera equipped with a 80-degree field of view wide angle and 100mm nine-degree field of view telephoto lens, attached to a 12-foot power/video cable. It produces a black-and-white 227 TV line signal scanned at 10 frames a second. Madrid, Goldstone and Honeysuckle Creek all will have equipment to make still photographs of the slow scan signal and to convert the signal to commercial TV format.

The color TV camera is a 12-pound Westinghouse camera with a zoom lens for close-up or wide angle use and a three-inch monitor which can be mounted on the camera or in the CM. It produces a standard 525-line, 30-frame-per-second signal in color by use of a rotating color wheel. The signal can be viewed in black and white. Only MSC, receiving the signal through Goldstone, will have equipment to colorize the signal.

Tentative planning is to use the color camera predominately, reverting to the black and white camera if there is difficulty with the color system but requiring at least one black and white transmission to Honeysuckle Creek. The following is a preliminary plan for TV passes based on a 12:49 May 18 launch:

GET	DATE/EDT	EVENT	
3:00 -3:15	18 - 3:48p	Transposition & dock	Madrid
3:15 -3:25	18 -	Transposition & dock	Goldstone
27:15 - 27 :25	19 - 4:03P	Translunar coast	Goldstone
54:00 - 51:10	20 - 6:48p	Translunar coast	Goldstone
72:20 - 72:35	21 - 1:08P	Pre-LOI-1	Goldstone/Madrid
80:45 - 80:53	21 - 9:33P	Post LOI-2	Goldstone
98:15-98.20	22- 3:01P	Post undock; formation	Goldstone
108:35-108:45	23 - 1:23a	APS Burn to Depletion	Goldstone
126:20-127:00	23 - 7:08P	Landmark Tracking	Goldstone
137:45-137:55	24 - 6:33a	Post-TEI	Honeysuckle*
152:35-152:45	24 - 9:23p	Transearth Coast	Goldstone
186:50-187:00	26 - 7:38a	Transearth coast	Goldstone

*Transmission from RCA black and white camera. All others planned to be from color camera.

APOLLO 10 PHOTOGRAPHIC TASKS

Still and motion pictures will be made of most spacecraft maneuvers as well as of the lunar surface and of crew activities in the Apollo 10 cabin.

The transposition, docking and lunar module ejection maneuver will be the first major event to be photographed in lunar orbit, the LM-active rendezvous sequence will be photographed from both the command and the lunar module.

During the period between the LM DPS phasing burn and the APS insertion burn, the commander and lunar module pilot will make still photos of the lunar ground track and of landing Site 2 from the eight-mile low point of the LM's flight path.

After rendezvous is complete and the LM APS depletion burn has been photographed, the crew will make stereo strip still photographs of the lunar surface and individual frames of targets of opportunity. Using the navigation sextant's optics as a camera lens system, lunar surface features and landmarks will be recorded on motion picture film. Additionally, the camera through-sextant system will photograph star-horizon and star-landmark combinations as they are superimposed in visual navigation sightings.

The Apollo 10 photography plan calls for motion pictures of crew activities such as intravehicular transfer through the CSM-LM docking tunnel and of other crew activities such as pressure suit donning.

Long-distance Earth and lunar terrain photographs will be shot with the 70mm still cameras,

Camera equipment carried on Apollo 10 consists of two 70mm Hasselblad, still cameras, each fitted with 80mm f/2.8 to f/22 Zeiss Planar lenses, a 250mm telephoto lens stowed aboard the command module, and associated equipment such as filters, ringsight, spotmeter and an intervalometer for stereo strip photography. One Hasselblad will be stowed in the LM and returned to the CSM after rendezvous. Hasselblad shutter speeds range from one second to 1/500 sec.

For motion pictures, two Maurer data acquisition cameras (one in the CSM, one in the LM) with variable frame speed selection will be used. Motion picture camera accessories include bayonet-mount lenses of 75, 18, and 5mm focal length, a right-angle mirror, a command module boresight bracket, a power cable, and an adapter for shooting through the sextant.

Apollo 10 film stowage includes six 70mm Hasselblad magazines — two exterior color reversal and four fine-grain black and white; and 12 140-foot 16mm magazines of motion picture film —- eight exterior color and four interior color— for a total of 1680 feet.

LUNAR DESCRIPTION

Terrain - Mountainous and crater-pitted, the former rising thousands of feet and the latter ranging from a few inches to 180 miles in diameter. The craters are thought to be formed by the impact of meteorites. The surface is covered with a layer of fine-grained material resembling silt or sand, as well as small rocks and boulders.

Environment - No air, no wind, and no moisture. The temperature ranges from 243 degrees in the two-week lunar day to 279 degrees below zero in the two-week lunar night. Gravity is one-sixth that of Earth. Micrometeoroids pelt the Moon (there is no atmosphere to burn them up). Radiation might present a problem during periods of unusual solar activity.

Dark Side - The dark or hidden side of the Moon no longer is a complete mystery. It was first photographed by a Russian craft and since then has been photographed many times, particularly by NASA's Lunar Orbiter spacecraft and Apollo 8.

Origin - There is still no agreement among scientists on the origin of the Moon. The three theories: (1) the Moon once was part of Earth and split off into its own orbit, (2) it evolved as a separate body at the same time as Earth, and (3) it formed elsewhere in space and wandered until it was captured by Earth's gravitational field.

Physical Facts

Diameter	2,160 miles (about ¼ that of Earth)
Circumference	6,790 miles (about ¼ that of Earth)
Distance from Earth	238,857 miles (mean; 221,463 minimum to 252,710 maximum)
Surface temperature	+243°F (Sun at zenith) -279°F (night)
Surface gravity	1/6 that of Earth
Mass	1/100th that of Earth
Volume	1/50th that of Earth
Lunar day and night	14 Earth days each
Mean velocity in orbit	2,287 miles per hour
Escape velocity	1.48 miles per second
Month (period of rotation around Earth)	27 days, 7 hours, 43 minutes

Apollo Lunar Landing Sites

Possible landing sites for the Apollo lunar module have been under study by NASA's Apollo Site Selection Board for more than two years. Thirty sites originally were considered. These have been narrowed down to four for the first lunar landing. (Site 1 currently not considered for first landing.)

Selection of the final five sites was based on high resolution photographs by Lunar Orbiter spacecraft, plus close-up photos and surface data provided by the Surveyor spacecraft which soft landed on the Moon.

The original sites are located on the visible side of the Moon within 45 degrees east and west of the Moon's center and 5 degrees north and south of its equator.

The final site choices were based on these factors:

* Smoothness (relatively few craters and boulders)

* Approach (no large hills, high cliffs, or deep craters that could cause incorrect altitude signals to the lunar module landing radar)

* Propellant requirements (selected sites require the least expenditure of spacecraft propellants)

* Recycle (selected sites allow effective launch preparation recycling if the Apollo Saturn V countdown is delayed)

* Free return (sites are within reach of the spacecraft launched on a free return translunar trajectory)

* Slope (there is little slope — less than 2 degrees in the approach path and landing area)

The Five Landing Sites Finally Selected Are:

Designations	Center Coordinates
Site 1	latitude 2° 37' 54" North longitude 34° 01' 31" East Site 1 is located on the east central part of the Moon in southeastern Mare Tranquillitatis. The site is approximately 62 miles (100 kilometers) east of the rim of Crater Maskelyne.
Site 2	latitude 0° 43' 56" North longitude 23° 38' 51" East Site 2 is located on the east central part of the Moon in southwestern Mare Tranquillitatis. The site is approximately 62 miles (100 kilometers) east of the rim of Crater Sabine and approximately 118 miles (190 kilometers) southwest of the Crater Maskelyne.
Site 3	latitude 0° 22' 27" North longitude 1° 20' 42" West Site 3 is located near the center of the visible face of the Moon in the southwestern part of Sinus Medii. The site is approximately 25 miles (40 kilometers) west of the center of the face and 21 miles (50 kilometers) southwest of the Crater Bruce.
Site 4	latitude 3° 38' 34" South longitude 36° 41' 53" West Site 4 is located on the west central part of the Moon in southeastern Oceanus Procellarum. The site is approximately 149 miles (240 kilometers) south of the rim of Crater Encke and 136 miles (220 kilometers) east of the rim of Crater Flamsteed.
Site 5	latitude 1° 46' 19" North longitude 41° 56' 20" West Site 5 is located on the west central part of the visible face in southeastern Oceanus Procellarum. The site is approximately 130 miles (210 kilometers) southeast of the rim of Crater Kepler and 118 miles (190 kilometers) north northeast of the rim of Crater Flamsteed.

APOLLO LUNAR LANDING SITES

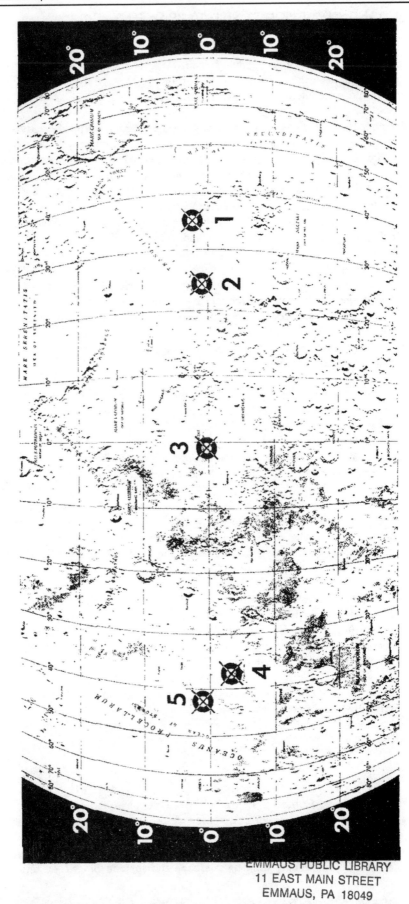

COMMAND AND SERVICE MODULE STRUCTURE, SYSTEMS

The Apollo spacecraft for the Apollo 10 mission is comprised of Command Module 106, Service Module 106, Lunar Module 4, a spacecraft-lunar module adapter (SLA) and a launch escape system. The SLA serves as a mating structure between the instrument unit atop the S-IVB stage of the Saturn V launch vehicle and as a housing for the lunar module.

Launch Escape System (LES) — Propels command module to safety in an aborted launch. It is made up of an open-frame tower structure, mounted to the command module by four frangible bolts, and three solid-propellant rocket motors: a 147,000 pound thrust launch escape system motor, a 2,400-pound-thrust pitch control motor, and a 31,500-pound-thrust tower jettison motor. Two canard vanes near the top deploy to turn the command module aerodynamically to an attitude with the heat-shield forward. Attached to the base of the launch escape tower is a boost protective cover composed of glass, cloth, and honeycomb, that protects the command module from rocket exhaust gases from the main and the jettison motors. The system is 33 feet tall, four feet in diameter at the base, and weighs 8,848 pounds.

Command Module (CM) Structure — The basic structure of the command module is a pressure vessel encased in heat shields, cone-shaped 11 feet 5 inches high, base diameter of 12 feet 10 inches, and launch weight 12,277 pounds. The command module consists of the forward compartment which contains two reaction control engines and components of the Earth landing system; the crew compartment or inner pressure vessel containing crew accommodations, controls and displays, and spacecraft systems; and the aft compartment housing ten reaction control engines and propellant tankage. The crew compartment contains 210 cubic feet of habitable volume.

Heat-shields around the three compartments are made of brazed stainless steel honeycomb with an outer layer of phenolic epoxy resin as an ablative material. Shield thickness, varying according to heat-loads, ranges from 0.7 inch at the apex to 2.7 inches at the aft end.

The spacecraft inner structure is of sheet-aluminum honeycomb bonded sandwich ranging in thickness from 0.25 inch thick at forward access tunnel to 1.5 inches thick at base.

CSM 106 and LM-4 are equipped with the probe-and-drogue docking hardware. The probe assembly is a folding coupling and impact attenuating device mounted on the CM tunnel that mates with a conical drogue mounted on the LM docking tunnel. After the docking latches are dogged down following a docking maneuver, both the probe and drogue assemblies are removed from the vehicle tunnels and stowed to allow free crew transfer between the CSM and LM.

Service Module (SM) Structure — The service module is a cylinder 12 feet 10 inches in diameter by 24 feet 7 inches high. For the Apollo 10 mission, it will weigh 51,371 pounds at launch. Aluminum honeycomb panels one inch thick form the outer skin, and milled aluminum radial beams separate the interior into six sections containing service propulsion system and reaction control fuel-oxidizer tankage, fuel cells, cryogenic oxygen and hydrogen, and onboard consumables.

Spacecraft-LM Adapter (SLA) Structure — The spacecraft LM adapter is a truncated cone 28 feet long tapering from 260 inches diameter at the base to 154 inches at the forward end at the service module mating line. Aluminum honeycomb 1.75 inches thick is the stressed-skin structure for the spacecraft adapter. The SLA weighs 4,000 pounds.

CSM Systems

Guidance, Navigation and Control System (GNCS) — Measures and controls spacecraft position, attitude and velocity, calculates trajectory, controls spacecraft propulsion system thrust vector, and displays abort data. The guidance system consists of three subsystems: inertial, made up of an inertial measurement unit and associated power and data components; computer which processes information to or from other components; and optics, including scanning telescope and sextant for celestial and/or landmark spacecraft

APOLLO SPACECRAFT

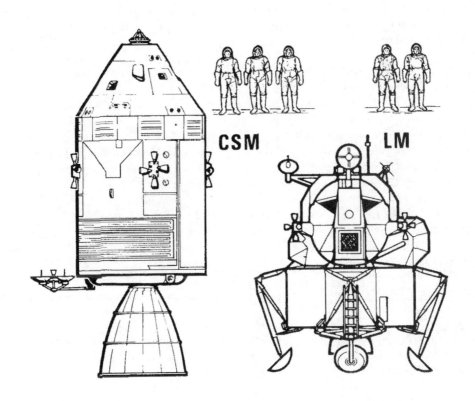

CSM

LM

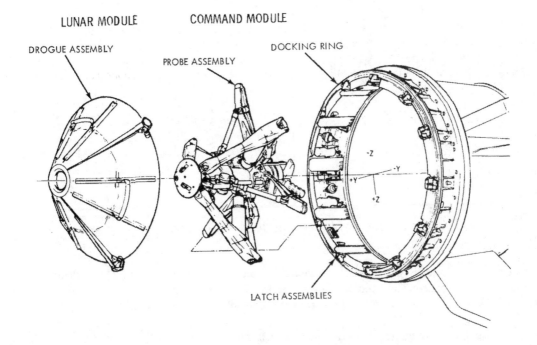

LUNAR MODULE   COMMAND MODULE

DROGUE ASSEMBLY

PROBE ASSEMBLY

DOCKING RING

LATCH ASSEMBLIES

APOLLO DOCKING MECHANISMS

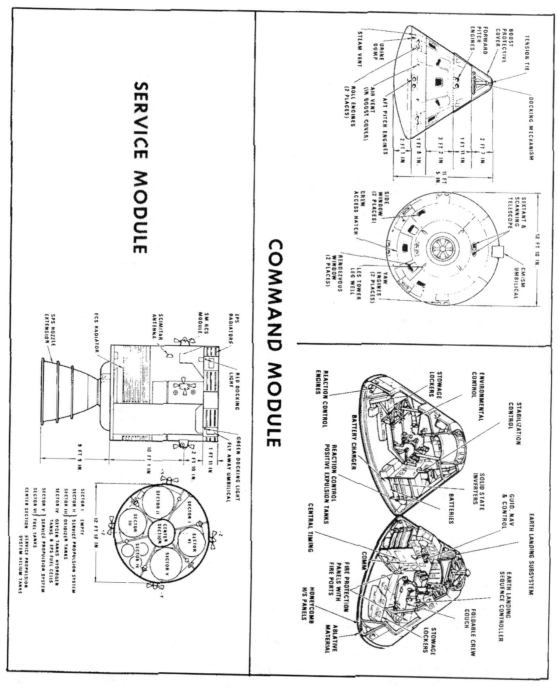

navigation. CSM 106 and subsequent modules are equipped with a VHF ranging device as a backup to the LM rendezvous radar.

Stabilization and Control System (SCS) — controls spacecraft rotation, translation, and thrust vector and provides displays for crew-initiated maneuvers; backs up the guidance system. It has three subsystems; attitude reference, attitude control, and thrust vector control.

Service Propulsion System (SPS) — Provides thrust for large spacecraft velocity changes through a gimbal-mounted 20,500 pound-thrust hypergolic engine using a nitrogen tetroxide oxidizer and a 50-50 mixture of unsymmetrical dimethyl hydrazine and hydrazine fuel. Tankage of this system is in the service module. The system responds to automatic firing commands from the guidance and navigation system or to manual commands from the crew. The engine provides a constant thrust rate. The stabilization and control system gimbals the engine to fire through the spacecraft center of gravity.

Reaction Control System (RCS) — The command module and the service module each has its own independent system. The SM RCS has four identical RCS "quads" mounted around the SM 90 degrees apart. Each quad has four 100 pound-thrust engines, two fuel and two oxidizer tanks and a helium pressurization sphere. The SM RCS provides redundant spacecraft attitude control through cross-coupling logic inputs from the stabilization and guidance systems.

Small velocity change maneuvers can also be made with the SM RCS. The CM RCS consists of two independent six-engine subsystems of six 93 pound-thrust engines each. Both subsystems are activated after CM separation from the SM: one is used for spacecraft attitude control during entry. The other serves in standby as a backup. Propellants for both CM and SM RCS are monomethyl hydrazine fuel and nitrogen tetroxide oxidizer with helium pressurization. These propellants are hypergolic, i.e. they burn spontaneously when combined without an igniter.

Electrical Power System (EPS) — Consists of three, 31 cell Bacon-type hydrogen-oxygen fuel cell power plants in the service module which supply 28-volt DC power, three 28-volt DC zinc-silver oxide main storage batteries in the command module lower equipment bay, and three 115-200 volt 400 hertz three phase AC inverters powered by the main 28-volt DC bus. The inverters are also located in the lower equipment bay. Cryogenic hydrogen and oxygen react in the fuel cell stacks to provide electrical power, potable water, and heat. The command module main batteries can be switched to fire pyrotechnics in an emergency. A battery charger restores selected batteries to full strength as required with power from the fuel cells.

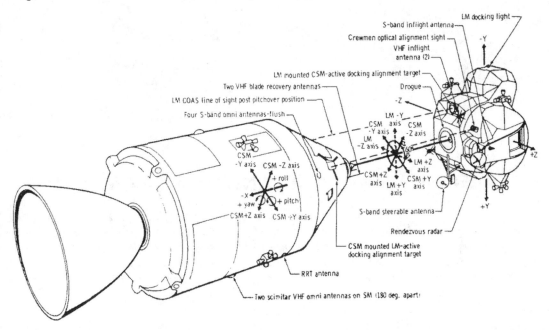

SPACECRAFT AXIS AND ANTENNA LOCATIONS

Environmental Control System (ECS) — Controls spacecraft atmosphere, pressure, and temperature and manages water. In addition to regulating cabin and suit gas pressure, temperature and humidity, the system removes carbon dioxide, odors and particles, and ventilates the cabin after landing. It collects and stores fuel cell potable water for crew use, supplies water to the glycol evaporators for cooling, and dumps surplus water overboard through the urine dump valve. Proper operating temperature of electronics and electrical equipment is maintained by this system through the use of the cabin heat exchangers, the space radiators, and the glycol evaporators.

Telecommunications System — Provides voice, television telemetry, and command data and tracking and ranging between the spacecraft and Earth, between the command module and the lunar module and between the spacecraft and the extravehicular astronaut. It also provides intercommunications between astronauts. The telecommunications system consists of pulse code modulated telemetry for relaying to Manned Space Flight Network stations data on spacecraft systems and crew condition, VHF/AM voice, and unified S-Band tracking transponders air-to-ground voice communications, onboard television, and a VHF recovery beacon. Network stations can transmit to the spacecraft such items as updates to the Apollo guidance computer and central timing equipment, and realtime commands for certain onboard functions.

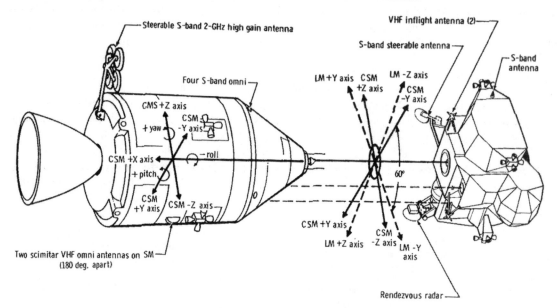

SPACECRAFT AXIS AND ANTENNA LOCATIONS

The high-gain steerable S-Band antenna consists of four, 31-inch-diameter parabolic dishes mounted on a folding boom at the aft end of the service module. Nested alongside the service propulsion system engine nozzle until deployment, the antenna swings out at right angles to the spacecraft longitudinal axis, with the boom pointing 52 degrees below the heads-up horizontal. Signals from the ground stations can be tracked either automatically or manually with the antenna's gimballing system. Normal S-Band voice and uplink/downlink communications will be handled by the omni and high-gain antennas.

Sequential System — Interfaces with other spacecraft systems and subsystems to initiate time critical functions during launch, docking maneuvers, sub-orbital aborts, and entry portions of a mission. The system also controls routine spacecraft sequencing such as service module separation and deployment of the Earth landing system.

Emergency Detection System (EDS) — Detects and displays to the crew launch vehicle emergency conditions, such as excessive pitch or roll rates or two engines out, and automatically or manually shuts down the booster and activates the launch escape system; functions until the spacecraft is in orbit.

Earth Landing System (ELS) — Includes the drogue and main parachute system as well as post-landing recovery aids. In a normal entry descent, the command module forward heat shield is jettisoned at 24,000 feet, permitting mortar deployment of two reefed 16.5-foot diameter drogue parachutes for orienting and decelerating the spacecraft. After disreef and drogue release three pilot mortar deployed chutes pull out the three main 83.3-foot diameter parachutes with two-stage reefing to provide gradual inflation in three steps. Two main parachutes out of three can provide a safe landing.

Recovery aids include the uprighting system, swimmer interphone connections, sea dye marker, flashing beacon, VHF recovery beacon, and VHF transceiver. The uprighting system consists of three compressor-inflated bags to upright the spacecraft if it should land in the water apex down (stable II position).

Caution and Warning System — Monitors spacecraft systems for out-of-tolerance conditions and alerts crew by visual and audible alarms so that crewmen may trouble-shoot the problem.

Controls and Displays — Provide readouts and control functions of all other spacecraft systems in the command and service modules. All controls are designed to be operated by crewmen in pressurized suits. Displays are grouped by system and located according to the frequency the crew refers to them.

LUNAR MODULE STRUCTURES, WEIGHT

The lunar module is a two-stage vehicle designed for space operations near and on the Moon. The LM is incapable of reentering the atmosphere. The lunar module stands 22 feet 11 inches high and is 31 feet wide (diagonally across landing gear).

Joined by four explosive bolts and umbilicals, the ascent and descent stages of the LM operate as a unit until staging, when the ascent stage functions as a single spacecraft for rendezvous and docking with the CSM.

Ascent Stage

Three main sections make up the ascent stage: the crew compartment, midsection, and aft equipment bay. Only the crew compartment and midsection are pressurized (4.8 psig; 337.4 gm/sq cm) as part of the LM cabin; all other sections of the LM are unpressurized. The cabin volume is 235 cubic feet (6.7 cubic meters). The ascent stage measures 12 feet 4 inches high by 14 feet 1 inch in diameter.

Structurally, the ascent stage has six substructural areas: crew compartment, midsection, aft equipment bay, thrust chamber assembly cluster supports, antenna supports and thermal and micrometeoroid shield.

The cylindrical crew compartment is a semimonocoque structure of machined longerons and fusion-welded aluminum sheet and is 92 inches (2.35 m) in diameter and 42 inches (1.07 m) deep. Two flight stations are equipped with control and display panels, armrests, body restraints, landing aids, two front windows, an overhead docking window, and an alignment optical telescope in the center between the two flight stations. The habitable volume is 160 cubic feet.

Two triangular front windows and the 32-inch (0.81 m) square inward-opening forward hatch are in the crew compartment front face.

External structural beams support the crew compartment and serve to support the lower interstage mounts at their lower ends. Ring-stiffened semimonocoque construction is employed in the midsection, with chem-milled aluminum skin over fusion-welded longerons and stiffeners. Fore-and-aft beams across the top of the midsection join with those running across the top of the cabin to take all ascent stage stress loads and, in effect, isolate the cabin from stresses.

The ascent stage engine compartment is formed by two beams running across the lower midsection deck and mated to the fore and aft bulkheads. Systems located in the midsection include the LM guidance computer, the power and servo assembly, ascent engine propellant tanks, RCS propellant tanks, the environmental control system, and the waste management section.

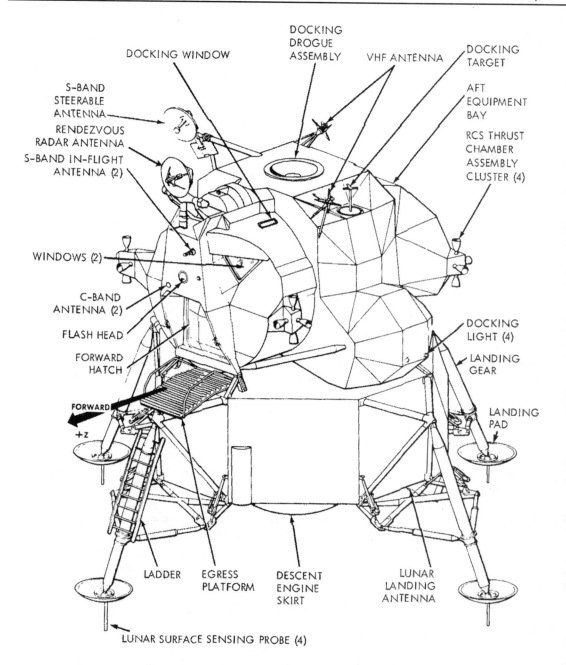

DOCKING WINDOW

DOCKING DROGUE ASSEMBLY

VHF ANTENNA

DOCKING TARGET

S-BAND STEERABLE ANTENNA

RENDEZVOUS RADAR ANTENNA

S-BAND IN-FLIGHT ANTENNA (2)

AFT EQUIPMENT BAY

RCS THRUST CHAMBER ASSEMBLY CLUSTER (4)

WINDOWS (2)

C-BAND ANTENNA (2)

FLASH HEAD

FORWARD HATCH

FORWARD

+Z

DOCKING LIGHT (4)

LANDING GEAR

LANDING PAD

LADDER   EGRESS PLATFORM   DESCENT ENGINE SKIRT

LUNAR LANDING ANTENNA

LUNAR SURFACE SENSING PROBE (4)

APOLLO LUNAR MODULE

A tunnel ring atop the ascent stage meshes with the command module latch assemblies. During docking, the ring and clamps are aligned by the LM drogue and the CSM probe.

The docking tunnel extends downward into the midsection 16 inches (40 cm). The tunnel is 32 inches (0.81 cm) in diameter and is used for crew transfer between the CSM and LM by crewmen. The upper hatch on the inboard end of the docking tunnel hinges downward and cannot be opened with the LM pressurized and undocked.

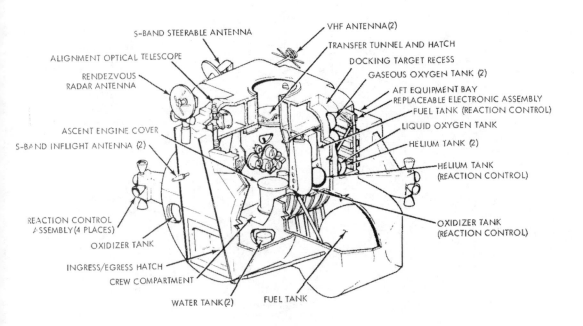

APOLLO LUNAR MODULE - ASCENT STAGE

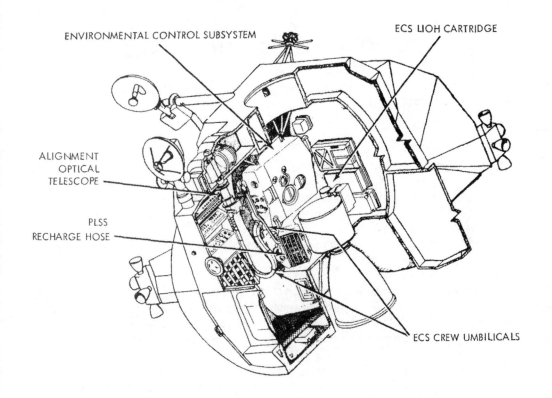

LM CABIN INTERIOR, LEFT HALF

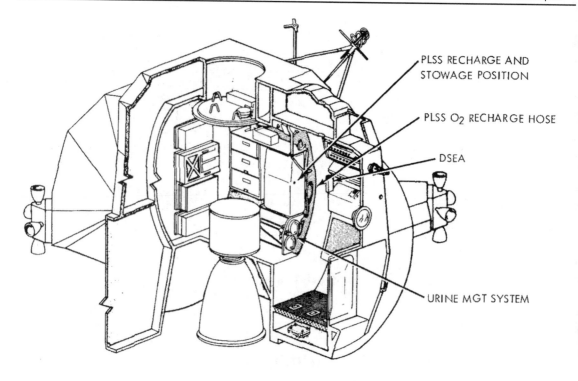

PLSS RECHARGE AND STOWAGE POSITION

PLSS O$_2$ RECHARGE HOSE

DSEA

URINE MGT SYSTEM

LM CABIN INTERIOR, RIGHT HALF

A thermal and micrometeoroid shield of multiple layers of mylar and a single thickness of thin aluminum skin encases the entire ascent stage structure.

Descent Stage

The descent stage consists of a cruciform load-carrying structure of two pairs of parallel beams, upper and lower decks, and enclosure bulkheads — all of conventional skin-and-stringer aluminum alloy construction. The center compartment houses the descent engine, and descent propellant tanks are housed in the four square bays around the engine. The descent stage measures 10 feet 7 inches high by 14 feet 1 inch in diameter.

Four-legged truss outriggers mounted on the ends of each pair of beams serve as SLA attach points and as "knees" for the landing gear main struts.

Triangular bays between the main beams are enclosed into quadrants housing such components as the ECS water tank, helium tanks, descent engine control assembly of the guidance, navigation and control subsystem, ECS gaseous oxygen tank, and batteries for the electrical power system. Like the ascent stage, the descent stage is encased in the mylar and aluminum alloy thermal and micrometeoroid shield.

The LM external platform, or "porch", is mounted on the forward outrigger just below the forward hatch. A ladder extends down the forward landing gear strut from the porch for crew lunar surface operations.

In a retracted position until after the crew mans the LM, the landing gear struts are explosively extended and provide lunar surface landing impact attenuation. The main struts are filled with crushable aluminum honeycomb for absorbing compression loads. Footpads 37 inches (0.95 m) in diameter at the end of each landing gear provide vehicle "flotation" on the lunar surface.

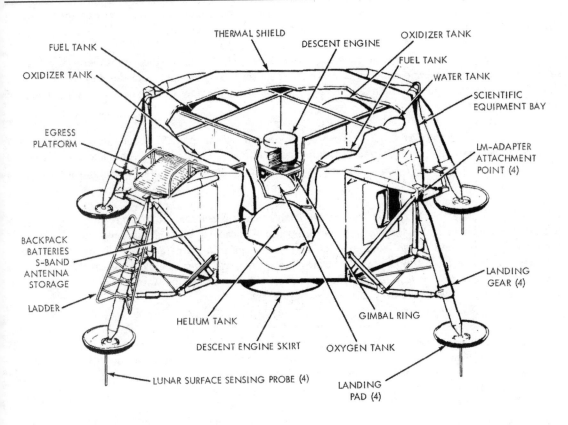

FUEL TANK

OXIDIZER TANK

THERMAL SHIELD   DESCENT ENGINE   OXIDIZER TANK

FUEL TANK

WATER TANK

SCIENTIFIC EQUIPMENT BAY

EGRESS PLATFORM

LM–ADAPTER ATTACHMENT POINT (4)

BACKPACK BATTERIES S-BAND ANTENNA STORAGE

LADDER

LANDING GEAR (4)

HELIUM TANK

GIMBAL RING

DESCENT ENGINE SKIRT   OXYGEN TANK

LUNAR SURFACE SENSING PROBE (4)   LANDING PAD (4)

Each pad is fitted with a lunar-surface sensing probe which signals the crew to shut down the descent engine upon contact with the lunar surface.

LM-4 flown on the Apollo 10 mission will have a launch weight of 30,849 pounds. The weight breakdown is as follows:

Ascent stage, dry	4,781 lbs.
Descent stage, dry	4,703 lbs.
RCS propellants	612 lbs.
DPS propellants	18,134 lbs.
APS propellants	2,619 lbs.
Total	30,849 lbs.

Lunar Module Systems

Electrical Power System — The LM DC electrical system consists of six silver zinc primary batteries four in the descent stage and two in the ascent stage, each with its own electrical control assembly (ECA). Power feeders from all primary batteries pass through circuit breakers to energize the LM DC buses, from which 28-volt DC power is distributed through circuit breakers to all LM systems. AC power (117v 400Hz) is supplied by two inverters, either of which can supply spacecraft AC load needs to the AC buses.

Environmental Control System — Consists of the atmosphere revitalization section, oxygen supply and cabin pressure control section, water management, heat transport section, and outlets for oxygen and water servicing of the Portable Life Support System (PLSS).

Components of the atmosphere revitalization section are the suit circuit assembly which cools and ventilates the pressure garments, reduces carbon dioxide levels, removes odors, noxious gases and excessive moisture;

the cabin recirculation assembly which ventilates and controls cabin atmosphere temperatures; and the steam flex duct which vents to space steam from the suit circuit water evaporator.

The oxygen supply and cabin pressure section supplies gaseous oxygen to the atmosphere revitalization section for maintaining suit and cabin pressure. The descent stage oxygen supply provides descent flight phase and lunar stay oxygen needs, and the ascent stage oxygen supply provides oxygen needs for the ascent and rendezvous flight phase.

Water for drinking, cooling, fire fighting, food preparation, and refilling the PLSS cooling water servicing tank is supplied by the water management section. The water is contained in three nitrogen-pressurized bladder-type tanks, one of 367-pound capacity in the descent stage and two of 47.5 pound capacity in the ascent stage.

The heat transport section has primary and secondary water-glycol solution coolant loops. The primary coolant loop circulates water-glycol for temperature control of cabin and suit circuit oxygen and for thermal control of batteries and electronic components mounted on cold plates and rails. If the primary loop becomes inoperative, the secondary loop circulates coolant through the rails and cold plates only. Suit circuit cooling during secondary coolant loop operation is provided by the suit loop water boiler. Waste heat from both loops is vented overboard by water evaporation or sublimators.

Communication System — Two S-band transmitter-receivers, two VHF transmitter-receivers, a signal processing assembly, and associated spacecraft antenna make up the LM communications system. The system transmits and receives voice, tracking and ranging data, and transmits telemetry data on 281 measurements and TV signals to the ground. Voice communications between the LM and ground stations is by S-band, and between the LM and CSM voice is on VHF.

Real-time commands to the lunar module are received and encoded by the digital uplink assembly — a black box tied in to the S-band receiver. The digital uplink assembly will be used on Apollo 10 to arm and fire the ascent propulsion system for the unmanned APS depletion burn following final docking and LM jettison. LM-4 will be the last spacecraft to be fitted with equipment for accepting real-time commands from the ground.

The data storage electronics assembly (DSEA) is a four channel voice recorder with timing signals with a 10-hour recording capacity which will be brought back into the CSM for return to Earth. DSEA recordings cannot be "dumped" to ground stations.

LM antennas are one 26-inch diameter parabolic S-band steerable antenna, two S-band inflight antennas and two VHF inflight antennas.

Guidance, Navigation and Control System — Comprised of six sections: primary guidance and navigation section (PGNS), abort guidance section (AGS), radar section, control electronics section (CES), and orbital rate drive electronics for Apollo and LM (ORDEAL).

* The PGNS is an inertial system aided by the alignment optical telescope, an inertial measurement unit, and the rendezvous and landing radars. The system provides inertial reference data for computations, produces inertial alignment reference by feeding optical sighting data into the LM guidance computer, displays position and velocity data, computes LM-CSM rendezvous data from radar inputs, controls attitude and thrust to maintain desired LM trajectory, and controls descent engine throttling and gimbaling

* The AGS is an independent backup system for the PGNS, having its own inertial sensor and computer.

* The radar section is made up of the rendezvous radar which provides CSM range and range rate, and line-of-sight angles for maneuver computation to the LM guidance computer; the landing radar which provide altitude and velocity data to the LM guidance computer during lunar landing. The rendezvous radar has an operating range from 80 feet to 400 nautical miles. The range transfer to the assembly, utilizing VHF electronics, is a passive responder to the CSM VHF ranging device and is a backup to the rendezvous radar.

* The CES controls LM attitude and translation about all axes. It also controls by PGNS command the automatic operation of the ascent and descent engines, and the reaction control thrusters. Manual attitude controller and thrust-translation controller commands are also handled by the CES.

* ORDEAL, displays on the flight director attitude indicator, is the computed local vertical in the pitch axis during circular, Earth or lunar orbits.

Reaction Control System — The LM has four RCS engine clusters of four 100 -pound (45.4 kg) thrust engines each which use helium pressurized hypergolic propellants. The oxidizer is nitrogen tetroxide, fuel is Aerozine 50 (50/50 blend of hydrazine and unsymmetrical dimethyl hydrazine). Propellant plumbing, valves and pressurizing components are in two parallel, independent systems, each feeding half the engines in each cluster. Either system is capable of maintaining attitude alone, but if one supply system fails, a propellant crossfeed allows one system to supply all 16 engines. Additionally, interconnect valves permit the RCS system to draw from ascent engine propellant tanks.

The engine clusters are mounted on outriggers 90 degrees apart on the ascent stage.

The RCS provides small stabilizing impulses during ascent and descent burns, controls LM attitude during maneuvers, and produces thrust for separation, and ascent/descent engine tank ullage. The system may be operated in either the pulse or steady-state modes.

Descent Propulsion System — Maximum rated thrust of the descent engine is 9,870 pounds (4,380.9 kg) and is throttleable between 1,050 pounds (476.7 kg) and 6,300 pounds (2,860.2 kg). The engine can be gimbaled six degrees in any direction for offset center of gravity trimming. Propellants are helium pressurized Aerozine 50 and nitrogen tetroxide.

Ascent Propulsion System — The 3,500-pound (1,589 kg) thrust ascent engine is not gimbaled and performs at full thrust. The engine remains dormant until after the ascent stage separates from the descent stage. Propellants are the same as are burned by the RCS engines and the descent engine.

Caution and Warning, Controls and Displays — These two systems have the same function aboard the lunar module as they do aboard the command module. (See CSM systems section.)

Tracking and Docking Lights — A flashing tracking light (once per second, 20 milliseconds duration) on the front face of the lunar module is an aid for contingency CSM-active rendezvous LM rescue. Visibility ranges from 400 nautical miles through the CSM sextant to 130 miles with the naked eye. Five docking lights analogous to aircraft running lights are mounted on the LM for CSM-active rendezvous: two forward yellow lights, aft white light port red light and starboard green light. All docking lights have about a 1,000-foot visibility.

SATURN V LAUNCH VEHICLE DESCRIPTION AND OPERATION

The Saturn V, 363 feet tall with the Apollo spacecraft in place, generates enough thrust to place a 125-ton payload into a 105-nm circular orbit of the Earth. It can boost about 50 tons to lunar orbit. The thrust of the three propulsive stages range from almost 7.6 million pounds for the booster to 230,000 pounds for the third stage at operating altitude. Including the instrument unit, the launch vehicle without the spacecraft is 281 feet tall.

First Stage

The first stage (S-IC) was developed jointly by the National Aeronautics and Space Administration's Marshall Space Plight Center, Huntsville, Ala. and the Boeing Co.

The Marshall Center assembled four S-IC stages: a structural test model, a static test version, and the first two flight stages. Subsequent flight stages are assembled by Boeing at the Michoud Assembly Facility, New Orleans. The S-IC stage destined for the Apollo 10 mission was the second flight booster static tested at the NASA-Mississippi Test Facility. The first S-IC test at MTF was on May 11, 1967, and the test of the second S-IC there — the booster for Apollo 10 — was completed Aug. 9, 1967. Earlier flight stages were static fired at the Marshall Center.

The S-IC stage boosts the space vehicle to an altitude of 35.8 nm at 50 nm downrange and increases the vehicle's velocity to 5,343 knots in 2 minutes 40 seconds of powered flight. It then separates and falls into the Atlantic Ocean about 351 nm downrange (30 degrees North latitude and 74 degrees West longitude) about nine minutes after liftoff.

Normal propellant flow rate to the five F-1 engines is 29,522 pounds per second. Four of the engines are mounted on a ring, each 90 degrees from its neighbor. These four are gimballed to control the rocket's direction of flight. The fifth engine is mounted rigidly in the center.

<u>Second Stage</u>

The second stage (S-II) like the third stage, uses high performance J-2 engines that burn liquid oxygen and liquid hydrogen. The stage's purpose is to provide stage boost nearly to Earth orbit.

SATURN V LAUNCH VEHICLE

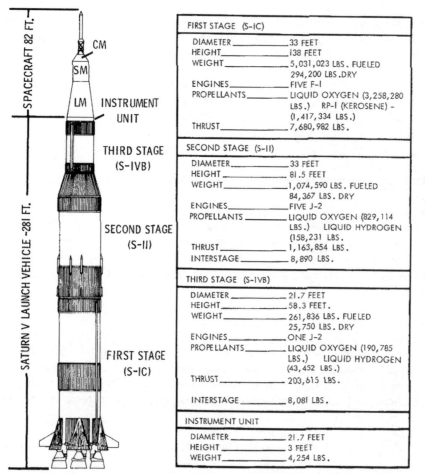

FIRST STAGE (S-IC)	
DIAMETER	33 FEET
HEIGHT	138 FEET
WEIGHT	5,031,023 LBS. FUELED 294,200 LBS. DRY
ENGINES	FIVE F-1
PROPELLANTS	LIQUID OXYGEN (3,258,280 LBS.) RP-1 (KEROSENE) - (1,417,334 LBS.)
THRUST	7,680,982 LBS.

SECOND STAGE (S-II)	
DIAMETER	33 FEET
HEIGHT	81.5 FEET
WEIGHT	1,074,590 LBS. FUELED 84,367 LBS. DRY
ENGINES	FIVE J-2
PROPELLANTS	LIQUID OXYGEN (829,114 LBS.) LIQUID HYDROGEN (158,231 LBS.)
THRUST	1,163,854 LBS.
INTERSTAGE	8,890 LBS.

THIRD STAGE (S-IVB)	
DIAMETER	21.7 FEET
HEIGHT	58.3 FEET.
WEIGHT	261,836 LBS. FUELED 25,750 LBS. DRY
ENGINES	ONE J-2
PROPELLANTS	LIQUID OXYGEN (190,785 LBS.) LIQUID HYDROGEN (43,452 LBS.)
THRUST	203,615 LBS.
INTERSTAGE	8,081 LBS.

INSTRUMENT UNIT	
DIAMETER	21.7 FEET
HEIGHT	3 FEET
WEIGHT	4,254 LBS.

NOTE: WEIGHTS AND MEASURES GIVEN ABOVE ARE FOR THE NOMINAL VEHICLE CONFIGURATION FOR APOLLO 10. THE FIGURES MAY VARY SLIGHTLY DUE TO CHANGES BEFORE LAUNCH TO MEET CHANGING CONDITIONS.

At outboard engine cutoff, the S-II separates and, following a ballistic trajectory, plunges into the Atlantic Ocean about 2,400 nm downrange from Kennedy Space Center (31 degrees North latitude and 34 degrees West longitude) about 20 minutes after liftoff.

Five J-2 engines power the S-II. The outer four engines are equally spaced on a 17.5-foot diameter circle. These four engines may be gimbaled through a plus or minus seven degree square pattern for thrust vector control. As on the first stage, the center engine (number 5) is mounted on the stage centerline and is fixed in position.

The S-II carries the rocket to an altitude of about 101.6 nm and a distance of some 888 nm downrange. Before burnout, the vehicle will be moving at a speed of 13,427 knots. The outer J-2 engines will burn 6 minutes 32 seconds during this powered phase, but the center engine will be cut off at 4 minutes 59 seconds of burn time.

The Space Division of North American Rockwell Corp. builds the S-II at Seal Beach, Calif. The cylindrical vehicle is made up of the forward skirt to which the third stage attaches, the liquid hydrogen tank, the liquid oxygen tank (separated from the hydrogen tank by a common bulkhead), the thrust structure on which the engines are mounted and an interstage section to which the first stage attaches. The common bulkhead between the two tanks is heavily insulated.

The S-II for Apollo 10 was static tested by North American Rockwell at the NASA Mississippi Test Facility on Aug. 9, 1968. This stage was shipped to the test site via the Panama Canal for the test firing.

Third Stage

The third stage (S-IVB) was developed by the McDonnell Douglas Astronautics Co. at Huntington Beach, Calif. At Sacramento, Calif., the stage passed a static firing test on Oct. 9, 1967 as part of the preparation for the Apollo 10 mission. The stage was flown directly to the NASA-Kennedy Space Center.

Measuring 58 feet 4 inches long and 21 feet 8 inches in diameter, the S-IVB weighs 25,750 pounds dry. At first ignition, it weighs 261,836 pounds. The interstage section weighs an additional 8,081 pounds. The stage's J-2 engine burns liquid oxygen and liquid hydrogen.

The stage provides propulsion twice during the Apollo 10 mission. The first burn occurs immediately after separation from the S-II. It will last long enough (156 seconds) to insert the vehicle and spacecraft into a circular Earth parking orbit at about 52 degrees West longitude and 32 degrees North latitude.

The second burn, which begins at 2 hours 33 minutes 25 seconds after liftoff (for first opportunity translunar injection) or 4 hours 2 minutes 5 seconds (for second TLI opportunity), will place the stage, instrument unit, and spacecraft into translunar trajectory. The burn will continue until proper TLI and conditions are met.

The fuel tanks contain 43,452 pounds of liquid hydrogen and 190,785 pounds of liquid oxygen at first ignition, totaling 234,237 pounds of propellants. Insulation between the two tanks is necessary because the liquid oxygen, at about 293 degrees below zero F is warm enough, relatively, to heat the liquid hydrogen, at 423 degrees below zero F, rapidly and cause it to turn into gas.

Instrument Unit

The Instrument unit (IU) is a cylinder three feet high and 21 feet 8 inches in diameter. It weighs 4,254 pounds and contains the guidance, navigation, and control equipment which will steer the vehicle through its Earth orbits and into the final translunar injection maneuver.

The IU also contains telemetry, communications, tracking, and crew safety systems, along with its own supporting electrical power and environmental control systems.

Components making up the "brain" of the Saturn V are mounted on cooling panels fastened to the inside surface of the instrument unit skin. The "cold plates" are part of a system that removes heat by circulating cooled fluid through a heat exchanger that evaporates water from a separate supply into the vacuum of space.

The six major systems of the instrument unit are structural, thermal control, guidance and control, measuring and telemetry, radio frequency, and electrical.

The instrument unit provides navigation, guidance, and control of the vehicle; measurement of vehicle performance and environment; data transmission with ground stations; radio tracking of the vehicle; checkout and monitoring of vehicle functions; initiation of stage functional sequencing; detection of emergency situations; generation and network distribution of electric power system operation; and preflight checkout and launch and flight operations.

A path-adaptive guidance scheme is used in the Saturn V instrument unit. A programmed trajectory is used in the initial launch phase with guidance beginning only after the vehicle has left the atmosphere. This is to prevent movements that might cause the vehicle to break apart while attempting to compensate for winds, jet streams, and gusts encountered in the atmosphere.

If such air currents displace the vehicle from the optimum trajectory in climb, the vehicle derives a new trajectory. Calculations are made about once each second throughout the flight. The launch vehicle digital computer and data adapter perform the navigation and guidance computations.

The ST-124M inertial platform — the heart of the navigation, guidance and control system —provides space-fixed reference coordinates and measures acceleration along the three mutually perpendicular axes of the coordinate system.

International Business Machines Corp., is prime contractor for the instrument unit and is the supplier of the guidance signal processor and guidance computer. Major suppliers of instrument unit components are: Electronic Communications, Inc., control computer; Bendix Corp., ST-124M inertial platform; and IBM Federal Systems Division, launch vehicle digital computer and launch vehicle data adapter.

Propulsion

The 41 rocket engines of the Saturn V have thrust ratings ranging from 72 pounds to more than 1.5 million pounds. Some engines burn liquid propellants, others use solids.

The five F-1 engines in the first stage burn RP-1 (kerosene) and liquid oxygen. Engines in the first stage develop approximately 1,536,197 pounds of thrust each at liftoff, building up to 1,822,987 pounds before cutoff. The cluster of five engines gives the first stage a thrust range from 7,680,982 million pounds at liftoff to 9,114,934 pounds just before center engine cutoff.

The F-1 engine weighs almost 10 tons, is more than 18 feet high and has a nozzle-exit diameter of nearly 14 feet. The F-1 undergoes static testing for an average 650 seconds in qualifying for the 160-second run during the Saturn V first stage booster phase. The engine consumes almost three tons of propellants per second.

The first stage of the Saturn V for this mission has eight other rocket motors. These are the solid-fuel retrorockets which will slow and separate the stage from the second stage. Each rocket produces a thrust of 87,900 pounds for 0.6 second.

The main propulsion for the second stage is a cluster of five J-2 engines burning liquid hydrogen and liquid oxygen. Each engine develops a mean thrust of more than 205,000 pounds at 5.0:1 mixture ratio (variable from 184,000 to 230,000 in phases of flight), giving the stage a total mean thrust of more than a million pounds.

Designed to operate in the hard vacuum of space, the 3,500-pound J-2 is more efficient than the F-1 because it burns the high-energy fuel hydrogen. F-1 and J-2 engines are produced by the Rocketdyne Division of North American Rockwell Corp.

The second stage has four 21,000-pound-thrust solid fuel rocket engines. These are the ullage rockets mounted on the S-IC/S-II interstage section. These rockets fire to settle liquid propellant in the bottom of the main tanks and help attain a "clean" separation from the first stage, they remain with the interstage when it drops away at second plane separation. Four retrorockets are located in the S-IVB aft interstage (which never separates from the S-II) to separate the S-II from the S-IVB prior to S-IVB ignition.

Eleven rocket engines perform various functions on the third stage. A single J-2 provides the main propulsive force; there are two jettisonable main ullage rockets and eight smaller engines in the two auxiliary propulsion system modules.

Launch Vehicle Instrumentation and Communication

A total of 2,342 measurements will be taken in flight on the Saturn V launch vehicles: 672 on the first stage, 986 on the second stage, 386 on the third stage, and 298 on the instrument unit.

The Saturn V has 16 telemetry systems: six on the first stage, six on the second stage, one on the third stage and three on the instrument unit. A C-band system and command system are also on the instrument unit. Each powered stage has a range safety system as on previous flights.

S-IVB Restart

The third stage of the Saturn V rocket for the Apollo 10 mission will burn twice in space. The second burn places the spacecraft on the translunar trajectory. The first opportunity for this burn is at 2 hours 33 minutes and 25 seconds after launch. The second opportunity for TLI begins at 4 hours 2 minutes and 5 seconds after liftoff.

The primary pressurization system of the propellant tanks for the S-IVB restart uses a helium heater. In this system, nine helium storage spheres in the liquid hydrogen tank contain gaseous helium charged to about 3,000 psi. This helium is passed through the heater which heats and expands the gas before it enters the propellant tanks. The heater operates on hydrogen and oxygen gas from the main propellant tanks.

The backup system consists of five ambient helium spheres mounted on the stage thrust structure. This system, controlled by the fuel repressurization control module, can repressurize the tanks in case the primary system fails. The restart will use the primary system. If that system fails, the backup system will be used.

The third stage for Apollo 10 will not be ignited for a third burn as on Apollo 9. Following spacecraft separation in translunar trajectory, the stage will undergo the normal J-2 engine chilldown sequence, stopping just short of reignition. On Apollo 10 there is no requirement for a third burn, and there will not be sufficient propellants aboard, most of the fuels having been expended during the translunar injection maneuver.

Differences in Apollo 9 and Apollo 10 launch Vehicles

Two modifications resulting from problems encountered during the second Saturn V flight were incorporated and proven successful on the third and fourth Saturn V missions. The new helium prevalve cavity pressurization system will again be flown on the first (S-IC) stage of Apollo 10, new augmented spark igniter lines which flew on the engines of the two upper stages of Apollo 8 and 9 will again be used on Apollo 10.

The major first stage (S-IC) differences between Apollo 9 and 10 are:

1. Dry weight was reduced from 295,600 to 294,200 pounds.

2. Weight at ground ignition increased from 5,026,200 to 5,031,023 pounds.

3. Instrumentation measurements were increased from 666 to 672.

S-II stage changes are:

1. Nominal vacuum thrust for J-2 engines increase will change maximum stage thrust from 1,150,000 to 1,168,694 pounds.

2. The approximate empty weight of the S-II has been reduced from 84,600 to 84,367 pounds. The S-IC/S-II interstage weight was reduced from 11,664 to 8,890 pounds.

3. Approximate stage gross liftoff weight was increased from 1,069,114 to 1,074,590 pounds.

4. Instrumentation measurements increased from 975 to 986.

Major differences on the S-IVB stage of Apollo 9 and 10 are:

1. S-IVB dry stage weight increased from 25,300 to 25,750 pounds. This does not include the 8,084-pound interstage section.

2. S-IVB gross stage weight at liftoff increased from 259,337 to 261,836 pounds.

3. Instrumentation measurements were increased from 296 to 386.

APOLLO 10 CREW

Life Support Equipment - Space Suits

Apollo 10 crewmen will wear two versions of the Apollo space suit: an intravehicular pressure garment assembly worn by the command module pilot and the extravehicular pressure garment assembly worn by the commander and the lunar module pilot. Both versions are basically identical except that the extravehicular version has an integral thermal/ meteoroid garment over the basic suit.

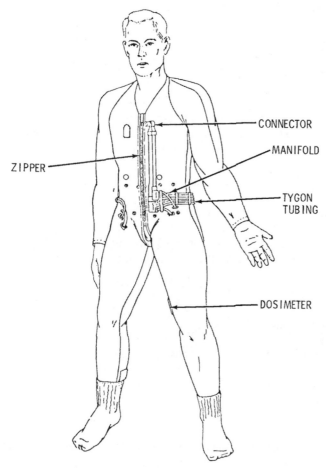

ZIPPER

CONNECTOR
MANIFOLD
TYGON TUBING
DOSIMETER

From the skin out, the basic pressure garment consists of a nomex comfort layer, a neoprene-coated nylon pressure bladder and a nylon restraint layer. The outer layers of the intravehicular suit are, from the inside out, nomex and two layers of Teflon-coated Beta cloth. The extravehicular integral thermal/meteoroid cover consists of a liner of two layers of neoprene-coated nylon, seven layers of Beta/Kapton spacer laminate, and an outer layer of Teflon-coated Beta fabric.

The extravehicular suit, together with a liquid cooling garment, portable life support system (PLSS), oxygen purge system, extravehicular visor assembly and other components make up the extravehicular mobility unit (EMU). The EMU provides an extravehicular crewman with life support for a four hour mission outside the lunar module without replenishing expendables. EMU total weight is 183 pounds. The intravehicular suit weighs 35.6 pounds.

Liquid cooling garment — A knitted nylon-spandex garment with a network of plastic tubing through which cooling water from the PLSS is circulated. It is worn next to the skin and replaces the constant wear-garment during EVA only.

Portable life support system — A backpack supplying oxygen at 3.9-psi and cooling water to the liquid cooling garment. Return oxygen is cleansed of solid and gas contaminants by a lithium hydroxide canister. The PLSS includes communications and telemetry equipment, displays and controls, and a main power supply. The PLSS is covered by a thermal insulation jacket. (One stowed in LM).

Oxygen purge system — Mounted atop the PLSS, the oxygen purge system provides a contingency 30-minute supply of gaseous oxygen in two two-pound bottles pressurized to 5,880 psia. The system may also be worn separately on the front of the pressure garment assembly torso. It serves as a mount for the VHF antenna for the PLSS. (Two stowed in LM).

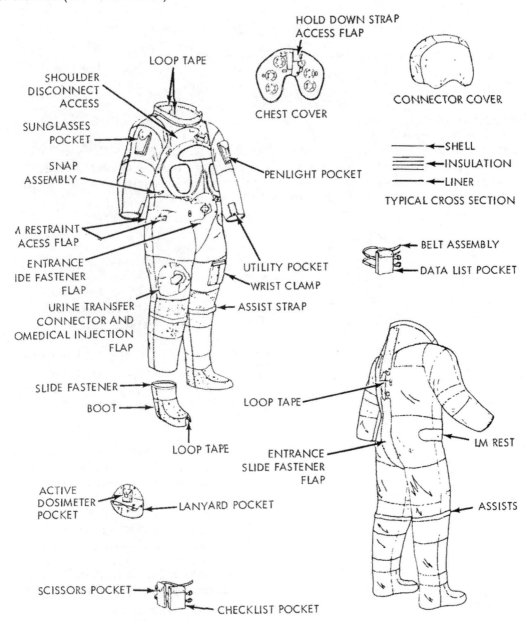

Extravehicular visor assembly — A polycarbonate shell and two visors with thermal control and optical coatings on them. The EVA visor is attached over the pressure helmet to provide impact, micrometeoroid, thermal and light protection to the EVA crewman.

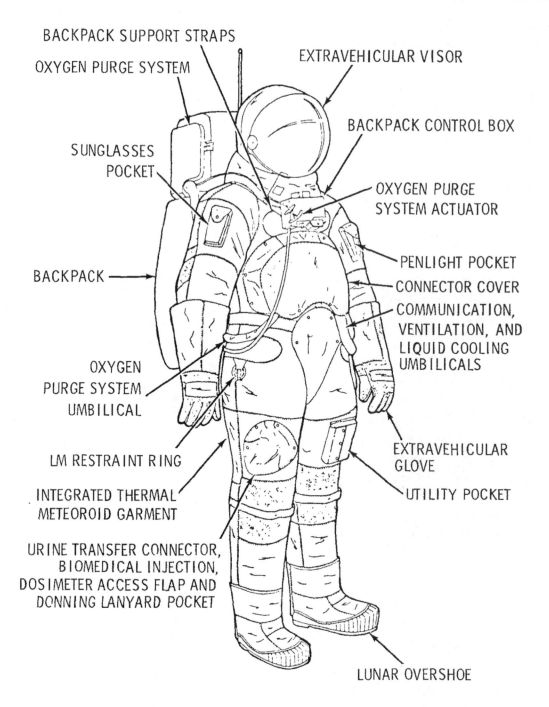

BACKPACK SUPPORT STRAPS

OXYGEN PURGE SYSTEM

EXTRAVEHICULAR VISOR

SUNGLASSES POCKET

BACKPACK CONTROL BOX

OXYGEN PURGE SYSTEM ACTUATOR

PENLIGHT POCKET

BACKPACK

CONNECTOR COVER

COMMUNICATION, VENTILATION, AND LIQUID COOLING UMBILICALS

OXYGEN PURGE SYSTEM UMBILICAL

LM RESTRAINT RING

EXTRAVEHICULAR GLOVE

INTEGRATED THERMAL METEOROID GARMENT

UTILITY POCKET

URINE TRANSFER CONNECTOR, BIOMEDICAL INJECTION, DOSIMETER ACCESS FLAP AND DONNING LANYARD POCKET

LUNAR OVERSHOE

Extravehicular gloves — Built of an outer shell of Chromel-R fabric and thermal insulation to provide protection when handling extremely hot and cold objects. The finger tips are made of silicone rubber to provide the crewman more sensitivity.

A one-piece constant-wear garment, similar to "long Johns", is worn as an undergarment for the space suit in intravehicular operations and for the inflight coveralls. The garment is porous-knit cotton with a waist-to-neck zipper for donning. Biomedical harness attach points are provided.

During periods out of the space suits, crewmen will wear two-piece Teflon fabric inflight coveralls for warmth and for pocket stowage of personal items.

Communications carriers ("Snoopy, hats") with redundant microphones and earphones are worn with the pressure helmet; a lightweight headset is worn with the inflight coveralls.

Meals

The Apollo 10 crew has a wide range of food items from which to select their daily mission space menu. More than 60 items comprise the food selection list of freeze-dried rehydratable foods. In addition, one "wet pack" meal-per-man per-day will be stowed for a total of 27. These meals, consisting of foil-wrapped beef and potatoes, ham and potatoes and turkey chunks and gravy, are similar to the Christmas meals carried aboard Apollo 8 and can be eaten with a spoon.

Water for drinking and rehydrating food is obtained from three sources in the command module — a dispenser for drinking water and two water spigots at the food preparation station, one supplying water at about 155 degrees F, the other at about 55 degrees F. The potable water dispenser squirts water continuously as long as the trigger is held down, and the food preparation spigots dispense water in one ounce increments.

Command module potable water is supplied from service module fuel cell byproduct water. Three one-pint "picnic jugs", or plastic bags, will be stowed aboard Apollo 10 for drinking water. Each crewman once a day will fill a bag with water and then spin it up to separate the suspended hydrogen gas from the water so that he will have hydrogenless water to drink the following day. The suspended hydrogen in the fuel cell byproduct water has caused intestinal discomfort to crewmen in previous Apollo missions.

A continuous-feed hand water dispenser similar to the one in the command module is used aboard the lunar module for cold-water rehydration of food packets stowed aboard the LM.

After water has been injected into a food bag, it is kneaded for about three minutes. The bag neck is then cut off and the food squeezed into the crewman's mouth. After a meal, germicide pills attached to the outside of the food bags are placed in the bags to prevent fermentation and gas formation. The bags are then rolled and stowed in waste disposal compartments.

The day-by-day, meal-by-meal Apollo 10 menu for each crewman for both the command module and the lunar module is listed on the following pages.

APOLLO 10 (STAFFORD)

MEAL	Day 1*, 5, 9	Day 2, 6,10	Day 3, 7, 11	Day 4, 8
A	Peaches Bacon Squares (8) Cinn Tstd Bread Cubes (4) Grapefruit Drink Orange Drink	Fruit Cocktail Sugar Coated Corn Flakes Bacon squares (8) Grapefruit Drink Grape Drink	Peaches Bacon Squares (8) Strawberry Cubes (4) Cocoa Orange Drink	Fruit Cocktail Sausage Patties Bacon Squares (8) Cocoa Grape Drink
B	Salmon Salad Chicken & Rice" Sugar Cookie Cubes (4) Cocoa Grape Punch	Potato Soup Chicken & Vegetables Tuna Salad Pineapple Fruitcake (4) Orange Drink	Cream of Chicken Soup (Turkey & Gravy - Wet Pack) Butterscotch Pudding Brownies (4) Grapefruit Drink	Potato Soup Pork & Scalloped Potatoes Applesauce Orange Drink
C	(Beef & Potatoes - Wet Pack) Cheese Cracker Cubes (4) Chocolate Pudding Orange-Grapefruit Drink	Spaghetti & Heat Sauce** (Ham & Potatoes - Wet Pack) Banana Pudding Pineapple-Grapefruit Drink	Pea Soup Beef Stew** Chicken Salad Chocolate Cubes Grape Punch	Shrimp Cocktail Chicken Stew** Turkey Bites (4) Date Fruitcake (4) Orange-Grapefruit Drink
CALORIES/DAY	2172	2179	2530	2010

*Day 1 consists of Meal C only
**New spoon-bowl package

APOLLO 10 (YOUNG)

MEAL	Day 1*, 5, 9	Day 2, 6, 10	Day 3, 7, 11	Day 4, 8
A	Peaches Bacon Squares (8) Cinn Tstd Bread Cubes (4) Grapefruit Drink Orange Drink	Fruit Cocktail Sugar Coated Corn Flakes Brownies (4) Grapefruit Drink Grape Drink	Peaches Bacon Squares (8) Strawberry Cubes (4) Cocoa Orange Drink	Fruit Cocktail Sausage Patties Bacon Squares (8) Cocoa Grape Drink
B	Salmon Salad Chicken & Rice" Sugar Cookie Cubes (4) Cocoa Grape Punch	Potato Soup Tuna Salad Chicken & Vegetables Pineapple Fruitcake (4) Pineapple-Grapefruit Drink	Cream of Chicken Soup (Turkey & Gravy - Wet Pack) Butterscotch Pudding Grapefruit Drink	Pea Soup Pork & Scalloped Potatoes Applesauce Orange Drink
C	(Beef & Potatoes - Wet Pack) Cheese Cracker Cubes (4) Chocolate Pudding Orange-Grapefruit Drink	Spaghetti & Meat Sauce** (Ham & Potatoes - Wet Pack) Banana Pudding Orange Drink	Beef Stew** Chicken Salad Corn Chowder Chocolate Cubes (4) Grape Punch	Shrimp Cocktail Chicken Stew** Turkey Bites (4) Date Fruitcake (4) Orange-Grapefruit Drink
CALORIES/DAY	2172	2145	2399	2020

*Day 1 consists of Meal C only
** New spoon-bowl package

APOLLO 10 (CERNAN)

MEAL	Day 1*, 5, 9	Day 2, 6, 10	Day 3, 7, 11	Day 4, 8
A	Peaches Bacon Squares (8) Cinn Tstd Bread Cubes (4) Orange Drink Orange-Pineapple Drink	Fruit Cocktail Sugar Coated Corn Flakes Bacon Squares (8) Orange Drink Grape Drink	Peaches Bacon Squares (8) Strawberry Cubes (4) Cocoa Orange Drink	Fruit Cocktail Sausage Patties Bacon Squares (8) Cocoa Grape Drink
B	Salmon Salad Chicken & Rice** Sugar Cookie Cubes (4) Cocoa Grape Punch	Potato Soup Tuna Salad Chicken & Vegetables Brownies (4) Orange-Grapefruit Drink	Cream of Chicken Soup (Turkey & Gravy - Wet Pack) Cinn Tstd Bread Cubes (4) Butterscotch Pudding Pineapple-Grapefruit Drink	Potato Soup Pork & Scalloped Potatoes Applesauce Orange Drink
C	Cream of Chicken Soup (Beef & Potatoes - Wet Pack) Cheese Cracker Cubes (4) Fruit Cocktail Orange-Grapefruit Drink	Spaghetti & Meat Sauce** (Ham & Potatoes - Wet Pack) Banana Pudding Orange Drink	Pea Soup Chicken Salad Beef Stew** Grape Punch	Shrimp Cocktail Chicken Stew** Turkey Bites (6) Chocolate Cubes (6) Orange-Grapefruit Drink
CALORIES/DAY	2026	2040	2298	2021

*Day 1 consists of Meal C only
** New spoon-bowl package

APOLLO 10 LM MENU

Day 1

Meal A

Fruit Cocktail
Bacon Squares (8)
Brownies (4)
Orange Drink
Grape Punch

Meal B

Beef and Vegetables
Pineapple Fruitcake (4)
Orange-Grapefruit Drink
Grape Punch

Meal C

Cream of Chicken Soup
Beef Hash
Strawberry Cubes (4)
Pineapple-Grapefruit Drink

Personal Hygiene

Crew personal hygiene equipment aboard Apollo 10 includes body cleanliness items, the waste management system and one medical kit.

Packaged with the food are a toothbrush and a two-ounce tube of toothpaste for each crewman. Each man-meal package contains a 3.5 by four inch wet-wipe cleansing towel. Additionally, three packages of 12 by 12 inch dry towels are stowed beneath the command module pilot's couch. Each package contains seven towels. Also stowed under the command module pilot's couch are seven tissue dispensers containing 53 three-ply tissues each.

Solid body wastes are collected in Gemini-type plastic defecation bags which contain a germicide to prevent bacteria and gas formation. The bags are sealed after use and stowed in empty food containers for post-flight analysis.

Urine collection devices are provided for use while wearing either the pressure suit or the inflight coveralls. The urine is dumped overboard through the spacecraft urine dump valve in the CM and stored in the LM.

The 5 x 5 x 8-inch medical accessory kit is stowed in a compartment on the spacecraft right side wall beside the lunar module pilot couch. The medical kit contains three motion sickness injectors, three pain suppression injectors, one two ounce bottle first aid ointments two one-ounce bottle eye drops, three nasal sprays, two compress bandages, 12 adhesive bandages, one oral thermometer and two spare crew biomedical harnesses. Pills in the medical kit are 60 antibiotic, 12 nausea, 12 stimulant, 18 pain killer, 60 decongestants, 24 diarrhea, 72 aspirin and 21 sleeping. Additionally, a small medical kit containing four stimulant, eight diarrhea, two sleeping and four pain killer pills, 12 aspirin, one bottle eye drops and two compress bandages is stowed in the lunar module flight data file compartment.

Survival Gear

The survival kit is stowed in two rucksacks in the right-hand forward equipment bay above the lunar module pilot.

Contents of rucksack No. I are: two combination survival lights, one desalter kit, three pair sunglasses, one radio beacon, one spare radio beacon battery and spacecraft connector cable, one knife in sheath, three water containers and two containers of Sun lotion.

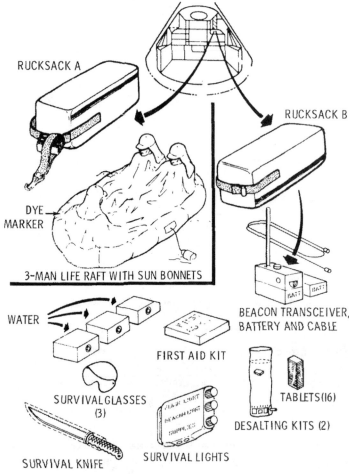

RUCKSACK A

DYE MARKER

3-MAN LIFE RAFT WITH SUN BONNETS

WATER

FIRST AID KIT

BEACON TRANSCEIVER, BATTERY AND CABLE

SURVIVAL GLASSES (3)

SURVIVAL KNIFE

SURVIVAL LIGHTS

TABLETS (16)

DESALTING KITS (2)

RUCKSACK B

Rucksack No. 2: one three-man life raft with CO2 inflater, one sea anchor, two sea dye markers, three sunbonnets, one mooring lanyard, three manlines, and two attach brackets.

The survival kit is designed to provide a 48-hour postlanding (water or land) survival capability for three crewmen between 40 degrees North and South latitudes.

Biomedical Inflight Monitoring

The Apollo 10 crew biomedical telemetry data received by the Manned Space Flight Network will be relayed for instantaneous display at Mission Control Center where heart rate and breathing rate data will be displayed on the flight surgeon's console. Heart rate and respiration rate average, range and deviation are computed and displayed on digital TV screens.

In addition, the instantaneous heart rate, real-time and delayed EKG and respiration are recorded on strip charts for each man.

Biomedical telemetry will be simultaneous from all crewmen while in the CSM, but selectable by a manual onboard switch in the LM.

Biomedical data observed by the flight surgeon and his team in the Life Support Systems Staff Support Room will be correlated with spacecraft and space suit environmental data displays.

Blood pressures are no longer telemetered as they were in the Mercury and Gemini programs. Oral temperature, however, can be measured onboard for diagnostic purposes and voiced down by the crew in case of inflight illness.

Rest-Work Cycles

All three Apollo 10 crewmen will sleep simultaneously during rest periods. The average mission day will consist of 16 hours of work and eight hours of rest. Two crewmen normally will sleep in the sleep stations (sleeping bags) under the couches, with the third man in the couch. During rest periods, one crewman will wear his communications headset.

The only exception to this sleeping arrangement will be during the rest period on lunar orbit insertion day, when two crewmen will sleep in the couches since the docking probe and drogue assemblies will be stowed in one of the sleep stations.

When possible, all three crewmen will eat together in one hour eat periods during which other activities will be held to a minimum.

Training

The crewmen of Apollo 10 have spent more than five hours of formal crew training for each hour of the lunar-orbit mission's eight-day duration. Almost 1,000 hours of training were in the Apollo 10 crew training syllabus over and above the normal preparations for the mission technical briefings and reviews, pilot meetings and study.

The Apollo 10 crewmen also took part in spacecraft manufacturing checkouts at the North American Rockwell plant in Downey, Calif., at Grumman Aircraft Engineering Corp., Bethpage, N.Y., and in prelaunch testing at NASA Kennedy Space Center. Taking part in factory and launch area testing has provided the crew with thorough operational knowledge of the complex vehicle.

Highlights of specialized Apollo 10 crew training topics are:

* Detailed series of briefings on spacecraft systems, operation and modifications.

* Saturn launch vehicle briefings on countdown, range safety, flight dynamics, failure modes and abort conditions. The launch vehicle briefings were updated periodically.

* Apollo Guidance and Navigation system briefings at the Massachusetts Institute of Technology Instrumentation Laboratory.

* Briefings and continuous training on mission photographic objectives and use of camera equipment.

* Extensive pilot participation in reviews of all flight procedures for normal as well as emergency situations.

* Stowage reviews and practice in training sessions in the spacecraft, mockups and command module simulators allowed the crewmen to evaluate spacecraft stowage of crew-associated equipment.

* More than 300 hours of training per man in command module and lunar module simulators at MSC and KSC, including closed loop simulations with flight controllers in the Mission Control Center. Other Apollo simulators at various locations were used extensively for specialized crew training.

* Entry corridor deceleration profiles at lunar-return conditions in the MSC Flight Acceleration Facility manned centrifuge.

* Zero-g aircraft flights using command module and lunar module mockups for EVA and pressure suit doffing/donning practice and training.

* Underwater zero-g training in the MSC Water Immersion Facility using spacecraft mockups to familiarize further crew with all aspects of CSM-LM docking tunnel intravehicular transfer and EVA in pressurized suits.

* Water egress training conducted in indoor tanks as well as in the Gulf of Mexico included uprighting from the Stable II position (apex down) to the Stable I position (apex up), egress onto rafts and helicopter pickup.

* Launch pad egress training from mockups and from the actual spacecraft on the launch pad for possible emergencies such as fire, contaminants and power failures.

* The training covered use of Apollo spacecraft fire suppression equipment in the cockpit.

* Planetarium reviews at Morehead Planetarium, Chapel Hill, N.C., and at Griffith Planetarium, Los Angeles, Calif., of the celestial sphere with special emphasis on the 37 navigational stars used by the Apollo guidance computer.

Crew Biographies

NAME: Thomas P. Stafford (Colonel, USAF) Apollo 10 commander NASA Astronaut

BIRTHPLACE AND DATE: Born September 17, 1930, in Weatherford, Okla., where his mother, Mrs. Mary Ellen Stafford, now resides.

PHYSICAL DESCRIPTION: Black hair, blue eyes; height- 6 feet; weight: 175 pounds.

EDUCATION: Graduated from Weatherford High School, Weatherford, Okla.; received a Bachelor of Science degree from the United States Naval Academy in 1952; recipient of an Honorary Doctorate of Science from Oklahoma City University in 1967.

MARITAL STATUS: Married to the former Faye L. Shoemaker of Weatherford, Okla. Her parents, Mr. and Mrs. Earle R. Shoemaker, reside in Thomas, Okla,

CHILDREN: Dionne, July 2, 1954; Karin, Aug. 28, 1957.

OTHER ACTIVITIES: His hobbies include handball, weight lifting and swimming.

ORGANIZATIONS: Member of the Society of Experimental Test pilots.

SPECIAL HONORS: Awarded two NASA Exceptional Service Medals and the Air Force Astronaut Wings; the Distinguished Flying Cross; the AIAA Astronautics Award; and co-recipient of the 1966 Harmon International Aviation Trophy.

EXPERIENCE: Stafford, an Air Force colonel, was commissioned in the United States Air Force upon graduation from Annapolis. Following his flight training, he flew fighter interceptor aircraft in the United States and Germany and later attended the USAF Experimental Flight Test School at Edwards Air Force Base, Calif.

He served as Chief of the Performance Branch at the USAF Aerospace Research Pilot School at Edwards and was responsible for the supervision and administration of the flying curriculum for student test pilots. He was also an instructor in flight test training and specialized academic subjects, establishing basic textbooks and directing the writing of flight test manuals for use by the staff and students. He is co-author of the Pilot's Handbook for Performance Flight Testing and the Aerodynamics Handbook for Performance Flight Testing.

He has accumulated over 5,000 hours flying time, of which over 4,000 hours are in jet aircraft.

CURRENT ASSIGNMENT: Colonel Stafford was selected as an astronaut by NASA in September 1962. He has since served as backup pilot for the Gemini 3 flight.

On Dec. 15, 1965, he and command pilot Walter M. Schirra were launched into space on the history-making Gemini 6 mission and subsequently participated in the first successful rendezvous of two manned maneuverable space craft by joining the already orbiting Gemini 7 crew. Gemini 6 returned to Earth on Dec. 16, 1965, after 25 hours 51 minutes and 24 seconds of flight.

He made his second flight as command pilot of the Gemini 9 mission. During this 3-day flight which began on June 3. 1966, the spacecraft attained a circular orbit of 161 statute miles; the crew performed three different types of rendezvous with the previously launched Augmented Target Docking Adapter; and pilot

Eugene Cernan logged two hours and ten minutes outside the spacecraft in extravehicular activity. The flight ended after 72 hours and 20 minutes with a perfect reentry and recovery as Gemini 9 landed within 0.4 nautical miles of the designated target point and 1½ miles from the prime recovery ship, USS WASP.

NAME: John W. Young (Commander, USN) Apollo 10 command module pilot NASA Astronaut

BIRTHPLACE AND DATE: Born in San Francisco Calif., on Sept. 24, 1930. His parents, Mr. and Mrs. William H. Young, reside in Orlando, Fla.

PHYSICAL DESCRIPTION- Brown hair; green eyes; height: 5 feet 9 inches; weight 165 pounds.

EDUCATION: Graduated from Orlando High School, Orlando, Fla.; received a Bachelor of Science degree in Aeronautical Engineering from the Georgia Institute of Technology in 1952.

MARITAL STATUS: Married to the former Barbara V. White of Savannah, Ga. Her parents, Mr. and Mrs. Robert A. White, reside in Jacksonville, Fla.

CHILDREN: Sandy, Apr. 30, 1957; John, Jan. 17, 1959.

OTHER ACTIVITIES: His hobbies are bicycle riding and handball.

ORGANIZATIONS: Member of the American institute of Aeronautic's and Astronautics and the Society of Experimental Test Pilots.

SPECIAL HONORS: Awarded two NASA Exceptional Service Medals, the Navy Astronaut Wings, and three Distinguished Flying Crosses.

EXPERIENCE: Upon graduation from Georgia Tech, Young entered the U.S. Navy in 1952 and holds the rank of commander.

He was a test pilot at the Naval Air Test Center from 1959 to 1962. Test projects included evaluations of the F8D and F4B fighter weapons systems. In 1962, he set world time-to-climb records to 3,000 and 25,000-meter altitudes in the F4B. Prior to his assignment to NASA he was Maintenance Officer of All-Weather-Fighter Squadron 143 at the Naval Air Station, Miramar, Calif.

He has logged more than 4,600 hours flying time, including more than 3,900 hours in jet aircraft.

CURRENT ASSIGNMENT: Commander Young was selected as an astronaut by NASA in September 1992.

He served as pilot on the first manned Gemini flight, a 3-orbit mission, launched on March 23, 1965, during which the crew accomplished the first manned spacecraft orbital trajectory modifications and lifting reentry, and flight tested all systems in Gemini 3. After this assignment, he was backup pilot for Gemini 6.

On July 18, 1966, Young occupied the command pilot seat for the Gemini 10 mission and, with Michael Collins as pilot, effected a successful rendezvous and docking with the Agena target vehicle. Then, they ignited the large Agena main engine to propel the docked combination to a record altitude of approximately 475 miles above the Earth— the first manned operation of a large rocket engine in space. They later performed a completely optical rendezvous (without radar) on a second passive Agena. After the rendezvous, while Young flew formation on the passive Agena, Collins performed extravehicular activity to it and recovered a micrometeorite detection experiment, accomplishing an in-space retrieval of the detector that had been orbiting the Earth for three months.

The flight was concluded after 3 days and 44 revolutions during which Gemini 10 traveled a total distance of 1,275,091 statute miles. Splashdown occurred in the West Atlantic, 529 statute miles east of Cape Kennedy, where Gemini 10 landed 2.6 miles from the USS GUADALCANAL within eye and camera range of the prime recovery vessel.

NAME: Eugene A. Cernan (Commanders USN) Apollo 10 lunar module pilot NASA Astronaut

BIRTHPLACE AND DATE: Born in Chicago, Ill., on March 14, 1934. His mother, Mrs. Andrew G. Cernan, resides in Bellwood, Ill.

PHYSICAL DESCRIPTION: Brown hair; blue eyes; height: 6 feet; weight: 170 pounds.

EDUCATION: Graduated from Proviso Township High School in Maywood, Ill.; received a Bachelor of Science degree in Electrical Engineering from Purdue University and a Master of Science degree in Aeronautical Engineering from the U.S. Naval Postgraduate School.

MARITAL STATUS: Married to the former Barbara J. Atchley of Houston, Tex.

CHILDREN: Teresa Dawn, March 4. 1963.

OTHER ACTIVITIES: His hobbies include gardening and all sports activities.

ORGANIZATIONS: Member of Tau Beta Pi, national engineering society; Sigma Xi, national science research society; and Phi Gamma Delta, national social fraternity.

SPECIAL HONORS: Awarded the NASA Exceptional Service Medal; the Navy Astronaut Wings; and the Distinguished Flying Cross.

EXPERIENCE: Cernan, a United States Navy commander, received his commission through the Navy ROTC program at Purdue. He entered flight training upon his graduation.

Prior to attending the Naval Postgraduate School, he was assigned to Attack Squadrons 126 and 113 at the Miramar, Calif., Naval Air Station.

He has logged more than 3,000 hours flying time with more than 2,810 hours in Jet aircraft.

CURRENT ASSIGNMENT: Commander Cernan was one of the third group of astronauts selected by NASA in October 1963.

He occupied the pilot seat alongside Command Pilot Tom Stafford on the Gemini 9 mission. During this 3 day flight which began on June 3, 1966, the spacecraft attained a circular orbit of 161 statute miles; the crew used three different techniques to effect rendezvous with the previously launched Augmented Target Docking Adapter, and Cernan logged two hours and ten minutes outside the spacecraft in extravehicular activity. The flight ended after 72 hours and 20 minutes with a perfect reentry and recovery as Gemini 9 landed within 1½ miles of the prime recovery ship USS WASP and 3/8 of a mile from the predetermined target point.

He has since served as backup pilot for Gemini 12.

APOLLO LAUNCH OPERATIONS

Prelaunch Preparations

NASA's John F. Kennedy Space Center performs preflight checkout, test, and launch of the Apollo 10 space vehicle. A government -Industry team of about 550 will conduct the final countdown from Firing Room 3 of the Launch Control Center (LCC).

The firing room team is backed up by more than 5,000 persons who are directly involved in launch operations at KSC from the time the vehicle and spacecraft stages arrive at the center until the launch is completed.

Initial checkout of the Apollo spacecraft is conducted in work stands and in the altitude chambers in the Manned Spacecraft Operations Building (MSOB) at Kennedy Space Center. After completion of checkout there, the assembled spacecraft is taken to the Vehicle Assembly Building (VAB) and mated with the launch vehicle. There the first integrated spacecraft and launch vehicle tests are conducted. The assembled space vehicle is then rolled out to the launch pad for final preparations and countdown to launch.

In mid-October 1968, flight hardware for Apollo 10 began arriving at Kennedy Space Center, just as Apollo 7 was being launched from Complex 34 on Cape Kennedy and as Apollo 8 and Apollo 9 were undergoing checkout at Kennedy Space Center.

The lunar module was the first piece of Apollo 10 flight hardware to arrive at KSC. The two stages were moved into the altitude chamber in the Manned Spacecraft Operations Building (MSOB) after an initial receiving inspection in October. In the chamber the LM underwent systems tests and both unmanned and manned chamber runs. During these runs the chamber air was pumped out to simulate the vacuum of space at altitudes in excess of 200,000 feet. There the spacecraft systems and the astronauts' life support systems were tested.

While the LM was undergoing preparation for its manned altitude chamber runs, the Apollo 10 command/service module arrived at KSC and after receiving inspection, it, too, was placed in an altitude chamber in the MSOB for systems tests and unmanned and manned chamber runs. The prime and back-up crews participated in the chamber runs on both the LM and the CSM.

In January, the LM and CSM were removed from the chambers. After installing the landing gear on the LM and the SPS engine nozzle on the CSM, the LM was encapsulated in the spacecraft LM adapter (SLA) and the CSM was mated to the SLA. On February 6, the assembled spacecraft was moved to the VAB where it was mated to the launch vehicle.

The launch vehicle flight hardware began arriving at KSC in late November, and by the end of December the three stages and the instrument unit were erected on the mobile launcher in high bay 2. This was the first time high bay 2, on the west side of the VAB, had been used for assembling a Saturn V. Tests were conducted on individual systems on each of the stages and on the overall launch vehicle before the spacecraft was erected atop the vehicle.

After spacecraft erection, the spacecraft and launch vehicle were electrically mated and the first overall test (plugs-in) of the space vehicle was conducted. In accordance with the philosophy of accomplishing as much of the checkout as possible in the VAB, the overall test was conducted before the space vehicle was moved to the launch pad.

The plugs-in test verified the compatibility of the space vehicle systems, ground support equipment, and off-site support facilities by demonstrating the ability of the systems to proceed through a simulated countdown, launch, and flight. During the simulated flight portion of the test, the systems were required to respond to both emergency and normal flight conditions.

The move to Pad B from the VAB on March 11 occurred while the Apollo 9 circled the Earth in the first manned test of the lunar module.

Apollo 10 will mark the first launch at Pad B on complex 39. The first two unmanned Saturn V launches and the manned Apollo 8 and 9 launches took place at Pad A. It also marked the first time that the transporter maneuvered around the VAB carrying a full load from high bay 2 on the 5-mile trip to the launch pad.

The space vehicle Flight Readiness Test was conducted in early April. Both the prime and backup crews participate in portions of the FRT, which is a final overall test of the space vehicle systems and ground support equipment when all systems are as near as possible to a launch configuration.

After hypergolic fuels were loaded aboard the space vehicle, and the launch vehicle first stage fuel (RP-1) was brought aboard, the final major test of the space vehicle began. This was the countdown demonstration test (CDDT) a dress rehearsal for the final countdown to launch. The CDDT for Apollo 10 was divided into a "wet" and a "dry" portion. During the first, or "wet" portion, the entire countdown, including propellant loading, was carried out down to T-8.9 seconds. The astronaut crews did not participate in the wet CDDT. At the completion of the wet CDDT, the cryogenic propellants (liquid oxygen and liquid hydrogen) were off-loaded, and the final portion of the countdown was re-run, this time simulating the fueling and with the prime astronaut crew participating as they will on launch day.

By the time Apollo 10 was entering the final phase of its checkout procedure at Complex 39B, crews had already started the checkout of Apollo 11 and Apollo 12. The Apollo 11 spacecraft completed altitude chamber testing and was mated to the launch vehicle in the VAB in mid-April as the Apollo 12 CSM and LM began checkout in the altitude chambers.

Because of the complexity involved in the checkout of the 363-foot-tall (110.6 meters) Apollo/Saturn V configuration, the launch teams make use of extensive automation in their checkout. Automation is one of the major differences in checkout used on Apollo compared to the procedures used in the Mercury and Gemini programs.

Computers, data display equipment, and digital data techniques are used throughout the automatic checkout from the time the launch vehicle is erected in the VAB through liftoff. A similar, but separate computer operation called ACE (Acceptance Checkout Equipment) is used to verify the flight readiness of the spacecraft. Spacecraft checkout is controlled from separate rooms in the Manned Spacecraft Operations Building.

LAUNCH COMPLEX 39

Launch Complex 39 facilities at the Kennedy Space Center were planned and built specifically for the Apollo Saturn V program, the space vehicle that will be used to carry astronauts to the Moon.

Complex 39 introduced the mobile concept of launch operations, a departure from the fixed launch pad techniques used previously at Cape Kennedy and other launch sites. Since the early 1950's when the first ballistic missiles were launched, the fixed launch concept had been used on NASA missions. This method called for assembly, checkout and launch of a rocket at one site — the launch pad. In addition to tying up the pad, this method also often left the flight equipment exposed to the outside influences of the weather for extended periods.

Using the mobile concept, the space vehicle is thoroughly checked in an enclosed building before it is moved to the launch pad for final preparations. This affords greater protection, a more systematic checkout process using computer techniques and a high launch rate for the future, since the pad time is minimal.

Saturn V stages are shipped to the Kennedy Space Center by ocean-going vessels and specially designed aircraft, such as the Guppy. Apollo spacecraft modules are transported by air. The spacecraft components are first taken to the Manned Spacecraft Operations Building for preliminary checkout. The Saturn V stages are brought immediately to the Vehicle Assembly Building after arrival at the nearby turning basin.

Apollo 10 is the first vehicle to be launched from Pad B, Complex 39 all previous Saturn V vehicles were launched Pad A at Complex 39. The historic first launch of the Saturn V, designated Apollo 4, took place Nov. 9. 1967 after a perfect countdown and on-time liftoff at 7 a.m. EST. The second Saturn V mission — Apollo 6—was conducted last April 4. The third Saturn V mission, Apollo 8. was conducted last Dec. 21-27. Apollo 9 was March 3-13, 1969.

The major components of complex 39 include: (1) the Vehicle Assembly Building (VAB) where the Apollo 10 was assembled and prepared; (2) the Launch Control Centers where the launch team conducts the preliminary checkout and final countdown; (3) the mobile launchers upon which the Apollo 10 was erected for checkout and from where it will be launched; (4) the mobile service structure, which provides external access to the space vehicle at the pad; (5) the transporter, which carries the space vehicle and mobile launcher, as well as the mobile service structure to the pad; (6) the crawlerway over which the space vehicle travels from the VAB to the launch pad; and (7) the launch pad itself.

<u>Vehicle Assembly Building</u>

The Vehicle Assembly Building is the heart of Launch Complex 39. Covering eight acres, it is where the 363-foot-tall space vehicle is assembled and tested.

The VAB contains 129,482,000 cubic feet of space. It is 716 feet long, and 518 feet wide and it covers 343,500 square feet of floor space.

The foundation of the VAB rests on 4,225 steel pilings, each 16 inches in diameter, driven from 150 to 170 feet to bedrock. If placed end to end, these pilings would extend a distance of 123 miles. The skeletal structure of the building contains approximately 60,000 tons of structural steel. The exterior is covered by more than a million square feet of insulated aluminum siding.

The building is divided into a high bay area 525 feet high and a low bay area 210 feet high, with both areas serviced by a transfer aisle for movement of vehicle stages.

The low bay work area, approximately 442 feet wide and 274 feet long, contains eight stage-preparation and checkout cells. These cells are equipped with systems to simulate stage interface and operation with other stages and the instrument unit of the Saturn V launch vehicle.

After the Apollo 10 launch vehicle upper stages arrived at the Kennedy Space Center, they were moved to the low bay of the VAB. Here, the second and third stages underwent acceptance and checkout testing prior to mating with the S-IC first stage atop mobile launcher 3 in the high bay area.

The high bay provides facilities for assembly and checkout of both the launch vehicle and spacecraft. It contains four separate bays for vertical assembly and checkout. At present, three bays are equipped, and the fourth will be reserved for possible changes in vehicle configuration.

Work platforms — some as high as three-story buildings — in the high bays provide access by surrounding the vehicle at varying levels. Each high bay has five platforms. Each platform consists of two bi-parting sections that move in from opposite sides and mate, providing a 360-degree access to the section of the space vehicle being checked.

A 10,000-ton-capacity air conditioning system, sufficient to cool about 3,000 homes, helps to control the environment within the entire office, laboratory, and workshop complex located inside the low bay area of the VAB. Air conditioning is also fed to individual platform levels located around the vehicle.

There are 141 lifting devices in the VAB, ranging from one ton hoists to two 250-ton high-lift bridge cranes.

The mobile launchers, carried by transporter vehicles, move in and out of the VAB through four doors in the high bay area, one in each of the bays. Each door is shaped like an inverted T. They are 152 feet wide and 114 feet high at the base, narrowing to 76 feet in width. Total door height is 456 feet.

The lower section of each door is of the aircraft hangar type that slides horizontally on tracks. Above this are seven telescoping vertical lift panels stacked one above the other, each 50 feet high and driven by an individual motor. Each panel slides over the next to create an opening large enough to permit passage of the mobile launcher.

<u>Launch Control Center</u>

Adjacent to the VAB is the Launch Control Center (LCC). This four-story structure is a radical departure from the dome-shaped blockhouses at other launch sites.

The electronic "brain" of Launch Complex 39, the LCC was used for checkout and test operations while Apollo 10 was being assembled inside the VAB. The LCC contains display, monitoring, and control equipment used for both checkout and launch operations.

The building has telemeter checkout stations on its second floor, and four firing rooms, one for each high bay of the VAB, on its third floor, three firing rooms contain identical sets of control and monitoring equipment, so that launch of a vehicle and checkout of others take place simultaneously. A ground computer facility is associated with each firing room.

The high speed computer data link is provided between the LCC and the mobile launcher for checkout of the launch vehicle. This link can be connected to the mobile launcher at either the VAB or at the pad.

The three equipped firing rooms have some 450 consoles which contain controls and displays required for the checkout process. The digital data links connecting with the high bay areas of the VAB and the launch pads carry vast amounts of data required during checkout and launch.

There are 15 display systems in each LCC firing room, with each system capable of providing digital information instantaneously.

Sixty television cameras are positioned around the Apollo/ Saturn V transmitting pictures on 10 modulated channels. The LCC firing room also contains 112 operational intercommunication channels used by the crews in the checkout and launch countdown.

Mobile Launcher

The mobile launcher is a transportable launch base and umbilical tower for the space vehicle. Three mobile launchers are used at Complex 39.

The launcher base is a two-story steel structure, 25 feet high, 160 feet long, and 135 feet wide. It is positioned on six steel pedestals 22 feet high when in the VAB or at the launch pad. At the launch pad, in addition to the six steel pedestals, four extendable columns also are used to stiffen the mobile launcher against rebound loads, if the Saturn engines cut off.

The umbilical tower, extending 398 feet above the launch platform, is mounted on one end of the launcher base. A hammerhead crane at the top has a hook height of 376 feet above the deck with a traverse radius of 85 feet from the center of the tower.

The 12-million-pound mobile launcher stands 445 feet high when resting on its pedestals. The base, covering about half an acre, is a compartment structure built of 25-foot steel girders.

The launch vehicle sits over a 45-foot-square opening which allows an outlet for engine exhausts into the launch pad trench containing a flame deflector. This opening is lined with a replaceable steel blast shield, independent of the structure, and is cooled by a water curtain initiated two seconds after liftoff.

There are nine hydraulically-operated service arms on the umbilical tower. These service arms support lines for the vehicle umbilical systems and provide access for personnel to the stages as well as the astronaut crew to the spacecraft.

On Apollo 10, one of the service arms is retracted early in the count. The Apollo spacecraft access arm is partially retracted at T-43 minutes. A third service arm is released at T-30 seconds, and a fourth at about T-16.5 seconds. The remaining five arms are set to swing back at vehicle first motion after T-0.

The service arms are equipped with a backup retraction system in case the primary mode fails.

The Apollo access arm (service arm 9), located at the 320 foot level above the launcher base, provides access to the spacecraft cabin for the closeout team and astronaut crews. The flight crew will board the spacecraft starting about T-2 hours, 40 minutes in the count. The access arm will be moved to a parked position, 12 degrees from the spacecraft, at about T-43 minutes. This is a distance of about three feet, which permits a rapid reconnection of the arm to the spacecraft in the event of an emergency condition. The arm is fully retracted at the T-5 minute mark in the count.

The Apollo 10 vehicle is secured to the mobile launcher by four combination support and hold-down arms mounted on the launcher deck. The hold-down arms are cast in one piece, about 6 x 9 feet at the base and 10 feet tall, weighing more than 20 tons. Damper struts secure the vehicle near its top.

After the engines ignite, the arms hold Apollo 10 for about six seconds until the engines build up to 95 percent thrust and other monitored systems indicate they are functioning properly. The arms release on receipt of a launch commit signal at the zero mark in the count. But the vehicle is prevented from accelerating too rapidly by controlled release mechanisms.

Transporter

The six-million-pound transporters, the largest tracked vehicles known, move mobile launchers into the VAB and mobile launchers with assembled Apollo space vehicles to the launch pad. They also are used to transfer the mobile service structure to and from the launch pads. Two transporters are in use at Complex 39.

The Transporter is 131 feet long and 114 feet wide. The vehicle moves on four double-tracked crawlers, each 10 feet high and 40 feet long, Each shoe on the crawler track is seven feet six inches in length and weighs about a ton.

Sixteen traction motors powered by four 1,000-kilowatt generators, which in turn are driven by two 2,750-horsepower diesel engines, provide the motive power for the transporter. Two 750 kW generators, driven by two 1,065-horsepower diesel engines, power the jacking, steering, lighting, ventilating and electronic systems.

Maximum speed of the transporter is about one-mile-per-hour loaded and about two-miles-per-hour unloaded, A five-mile trip to Pad B with a mobile launcher, made at less than maximum speed, takes approximately 10-12 hours.

The transporter has a leveling system designed to keep the top of the space vehicle vertical within plus-or-minus 10 minutes of arc — about the dimensions of a basketball.

This system also provides leveling operations required to negotiate the five percent ramp which leads to the launch pad and keeps the load level when it is raised and lowered on pedestals both at the pad and within the VAB.

The overall height of the transporter is 20 feet from ground level to the top deck on which the mobile launcher is mated for transportation. The deck is flat and about the size of a baseball diamond (90 by 90 feet).

Two operator control cabs, one at each end of the chassis located diagonally opposite each other, provide totally enclosed stations from which all operating and control functions are coordinated.

Crawlerway

The transporter moves on a roadway 131 feet wide, divided by a median strip. This is almost as broad as an eight-lane turnpike and is designed to accommodate a combined weight of about 18 million pounds.

The roadway is built in three layers with an average depth of seven feet. The roadway base layer is two-and-one-half feet of hydraulic fill compacted to 95 percent density. The next layer consists of three feet of crushed rock packed to maximum density, followed by a layer of one foot of selected hydraulic fill. The bed is topped and sealed with an asphalt prime coat.

On top of the three layers is a cover of river rock, eight inches deep on the curves and six inches deep on the straightway. This layer reduces the friction during steering and helps distribute the load on the transporter bearings.

Mobile Service Structure

A 402-foot-tall, 9.8-million-pound tower is used to service the Apollo launch vehicle and spacecraft at the pad. The 40-story steel-trussed tower, called a mobile service structure, provides 360-degree platform access to the Saturn launch vehicle and the Apollo spacecraft,

The service structure has five platforms — two self-propelled and three fixed, but movable. Two elevators carry personnel and equipment between work platforms. The platforms can open and close around the 363-foot space vehicle.

After depositing the mobile launcher with its space vehicle on the pad, the transporter returns to a parking area about 13,000 feet from pad B. There it picks up the mobile service structure and moves it to the launch pad. At the pad, the huge tower is lowered and secured to four mount mechanisms.

The top three work platforms are located in fixed positions which serve the Apollo spacecraft. The two lower movable platforms serve the Saturn V.

The mobile service structure remains in position until about T-11 hours when it is removed from its mounts and returned to the parking area.

Water Deluge System

A water deluge system will provide a million gallons of industrial water for cooling and fire prevention during launch of Apollo 10. Once the service arms are retracted at liftoff, a spray system will come on to cool these arms from the heat of the five Saturn F-1 engines during liftoff.

On the deck of the mobile launcher are 29 water nozzles. This deck deluge will start immediately after liftoff and will pour across the face of the launcher for 30 seconds at the rate of 50,000 gallons-per-minute. After 30 seconds, the flow will be reduced to 20,000 gallons-per-minute.

Positioned on both sides of the flame trench are a series of nozzles which will begin pouring water at 8,000 gallons-per-minute, 10 seconds before liftoff. This water will be directed over the flame deflector.

Other flush mounted nozzles, positioned around the pad, will wash away any fluid spill as a protection against fire hazards.

Water spray systems also are available along the egress route that the astronauts and closeout crews would follow in case an emergency evacuation was required.

Flame Trench and Deflector

The flame trench is 58 feet wide and approximately six feet above mean sea level at the base. The height of the trench and deflector is approximately 42 feet.

The flame deflector weighs about 1.3 million pounds and is stored outside the flame trench on rails. When it is moved beneath the launcher, it is raised hydraulically into position. The deflector is covered with a four-and-one-half-inch thickness of refractory concrete consisting of a volcanic ash aggregate and a calcium aluminate binder. The heat and blast of the engines are expected to wear about three-quarters of an inch from this refractory surface during the Apollo 10 launch.

Pad Areas

Both Pad A and Pad B of Launch Complex 39 are roughly octagonal in shape and cover about one fourth of a square mile of terrain.

The center of the pad is a hardstand constructed of heavily reinforced concrete. In addition to supporting the weight of the mobile launcher and the Apollo Saturn V vehicle, it also must support the 9.8-million-pound mobile service structure and 6-million-pound transporter, all at the same time. The top of the pad stands some 48 feet above sea level.

Saturn V propellants — liquid oxygen, liquid hydrogen and RP-1 are stored near the pad perimeter.

Stainless tools vacuum-jacketed pipes carry the liquid oxygen (LOX) and liquid hydrogen from the storage tanks to the pad, up the mobile launcher, and finally into the launch vehicle propellant tanks.

LOX is supplied from a 900,000-gallon storage tank. A centrifugal pump with a discharge pressure of 320 pounds per-square-inch pumps LOX to the vehicle at flow rates as high as 10,000-gallons-per-minute.

Liquid hydrogen, used in the second and third stages, is stored in an 850,000-gallon tank, and is sent through 1,500 feet of 10-inch, vacuum-jacketed invar pipe. A vaporizing heat exchanger pressurizes the storage tank to 60 psi for a 10,000 gallons-per-minute, flow rate.

The RP-1 fuel, a high grade of kerosene is stored in three tanks — each with a capacity of 86,000 gallons. It is pumped at a rate of 2,000 gallons-per-minute at 175 psig.

The complex 39 pneumatic system includes a converter compressor facility a pad high-pressure gas storage battery, a high-pressure storage battery in the VAB, low and high-pressure cross-country supply lines, high-pressure hydrogen storage and conversion equipment and pad distribution piping to pneumatic control panels. The various purging systems require 187,000 pounds of liquid nitrogen and 21,000 gallons of helium.

Pad B is virtually a twin of Pad A, The top of Pad B is 5 feet higher in elevation above mean sea level than Pad A to provide better drainage of the general area plus better drainage from holding and burn ponds.

The electrical substation for Pad B is located underneath the west slope of the pad whereas the corresponding substation for Pad A is in the open approximately 150 feet from the lower edge of the west slope of the pad. The pad B design change was made to harden the substation against the launch environment. The only other major difference is in the location of the industrial/ fire/ potable water valve pit. At Pad A it's on the west side of the Pad and at Pad B it's on the east side of the pad. The difference rests in the routing of water lines alongside the crawlerway.

Basic construction work on Pad B began on Dec. 7, 1964, and the facility was accepted by the government on August 22, 1966. The intervening period has been spent in equipping the pad and bringing it up to launch readiness.

Mission Control Center

The Mission Control Center at the Manned Spacecraft Center, Houston, is the focal point for Apollo flight control activities. The center receives tracking and telemetry data from the Manned Space Flight Network, processes this data through the Mission Control Center Real-Time Computer Complex, and displays this data to the flight controllers and engineers in the Mission Operations Control Room and staff support rooms.

The Manned Space Flight Network tracking and data acquisition stations link the flight controllers at the center to the spacecraft.

For Apollo 10 all network stations will be remote sites, that is, without flight control teams. All uplink commands and voice communications will originate from Houston, and telemetry data will be sent back to Houston at high speed rates (2,400 bits-per-second), on two separate data lines. They can be either real time or playback information.

Signal flow for voice circuits between Houston and the remote sites is via commercial carriers usually satellite, wherever possible using leased lines which are part of the NASA Communications Network.

Commands are sent from Houston to NASA's Goddard Space Flight Center, Greenbelt, Md., on lines which link computers at the two points. The Goddard communication computers provide automatic switching facilities and speed buffering for the command data. Data are transferred from Goddard to remote sites on high speed (2,400 bits-per-second) lines. Command loads also can be sent by teletype from Houston to the remote sites at 100 words-per-minute. Again, Goddard computers provide storage and switching functions.

Telemetry data at the remote site are received by the RF receivers, processed by the pulse code modulation ground stations, and transferred to the 642B remote-site telemetry computer for storage. Depending on the format selected by the telemetry controller at Houston, the 642B will send the desired format through a 2010 data transmission unit which provides parallel to serial conversion, and drives a 2,400 bit-per-second mode.

The data mode converts the digital serial data to phase-shifted keyed tones which are fed to the high speed data lines of the communications network.

Tracking data are sent from the sites in a low speed (100 words) teletype format and a 240-bit block high speed (2,400 bits) format. Data rates are one sample-6 seconds for teletype and 10 samples (frames) per second for high speed data.

All high-speed data, whether tracking or telemetry, which originate at a remote site are sent to Goddard on high speed lines. Goddard reformats the data when necessary and sends them to Houston in 600-bit blocks at a 40,800 bits-per second rate. Of the 600-bit block, 480 bits are reserved for data, the other 120 bits for address, sync, intercomputer instructions, and polynomial error encoding.

All wideband 40,800 bits-per-second data originating at Houston are converted to high speed (2,400 bits-per-second) data at Goddard before being transferred to the designated remote site.

MANNED SPACE FLIGHT NETWORK

The Manned Space Flight Network (MSFN) will support the complete Apollo spacecraft, operating at lunar distance, for the first time in Apollo 10. The network had its initial service with lunar distances in Apollo 8 last December, but that flight did not carry the lunar module.

For Apollo 10, the MSFN will employ 17 ground stations (including three wing, or backup, sites), four instrumented ships and six to eight instrumented aircraft, to track spacecraft position and furnish a large volume of communications, television and telemetry services.

Essentially, the entire network is designed to provide reliable and continuous communications with the astronauts, launch vehicle and spacecraft from liftoff through lunar orbit to splashdown. It will keep ground controllers in close contact with the spacecraft and astronauts at all times, except for approximately 45 minutes when Apollo 10 will be behind the Moon during each lunar orbit and the time between stations while in Earth orbit.

As the space vehicle lifts off from Kennedy Space Center, the tracking stations will be watching it. As the

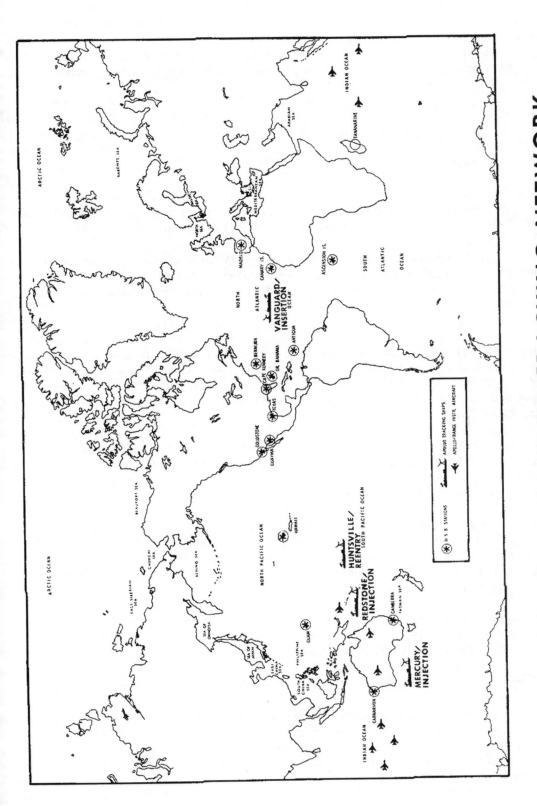

MANNED SPACE FLIGHT TRACKING NETWORK

Saturn ascends, voice and data will be instantaneously transmitted to Mission Control Center (MCC) in Houston. Data will be run through computers at MCC for visual display to flight controllers.

Depending on the launch azimuth, a string of 30-foot diameter antennas around the Earth will keep tabs on Apollo 10 and transmit information back to Houston: beginning with the station at Merritt Island, Fla.; thence Grand Bahama Island, Bermuda; the tracking ship Vanguard; Canary Island; Carnarvon, Australia; Hawaii, tracking ship Redstone, Guaymas, Mexico; and Corpus Christi, Tex.

To inject Apollo 10 into translunar trajectory MCC will send a signal through one of the land stations or one of the Apollo ships in the Pacific. As the spacecraft head for the Moon, the engine burn will be monitored by the ships and an Apollo Range Instrumentation Aircraft (ARIA). The ARIA provides a relay for the astronauts' voices and data communication with Houston.

As the spacecraft moves away from Earth, the smaller 30-foot diameter antennas communicate first with the spacecraft. At a spacecraft altitude of 10,000 miles the tracking function goes to the more powerful 85-foot antennas. These are located near Madrid, Spain; Goldstone, Calif.; and Canberra, Australia.

The 85-foot antennas are spaced at approximately 120 degree intervals around Earth so at least one antenna has the Moon in view at all times. As the Earth revolves from west to east, one station hands over control to the next station as it moves into view of the spacecraft. In this way, continuous data and communication flow is maintained.

Data are constantly relayed back through the huge antennas and transmitted via the NASA Communications Network (NASCOM) a half million miles of land and underseas cables and radio circuits, including those through communications satellites, to MCC. This information is fed into computers for visual display in Mission Control. For example, a display would show the exact position of the spacecraft on a large map. Returning data could indicate a drop in power or some other difficulty which would result in a red light going on to alert a flight controller to corrective action.

Returning data flowing to the Earth stations give the necessary information for commanding mid-course maneuvers to keep the Apollo 10 in a proper trajectory for orbiting the Moon. While the flight is in the vicinity of the Moon, these data indicate the amount of retrograde burn necessary for the service module engine to place the spacecraft units in lunar orbit.

Once the lunar module separates from the command module/ service module and goes into a separate lunar orbit, the MSFN will be required to keep track of both craft and provide continuous two-way communications and telemetry between them and the Earth. The prime antenna at each of the three MSFN deep space tracking stations will handle one craft while the wing or back-up antenna at each of these stations will handle the other craft during each pass.

Continuous tracking and acquisition of data between Earth and the Apollo spacecraft will provide support for the Apollo rendezvous and docking maneuvers. This information also will be used to determine the time and duration of the service module propulsion engine burn required to place the command/ service module into a precise trajectory for reentering the Earth's atmosphere at the planned location. As the spacecraft moves toward Earth at about 25,000 miles-per-hour, it must reenter at the proper angle.

Data coming to the various tracking stations and ships are fed into the computers at MCC. From computer calculations, the flight controllers will provide the returning spacecraft with the necessary information to make an accurate reentry. Appropriate MSFN stations, including tracking ships and aircraft positioned in the Pacific for this event are on hand to provide support during reentry. An ARIA aircraft will relay astronaut voice communications to MCC and antennas on reentry ships will follow the spacecraft.

During the journey to the Moon and back, television will be received from the spacecraft at the three 85-foot antennas around the world, in Spain, California, and Australia. Scan converters permit immediate transmission of commercial quality television via NASCOM to Mission Control where it will be released to TV networks,

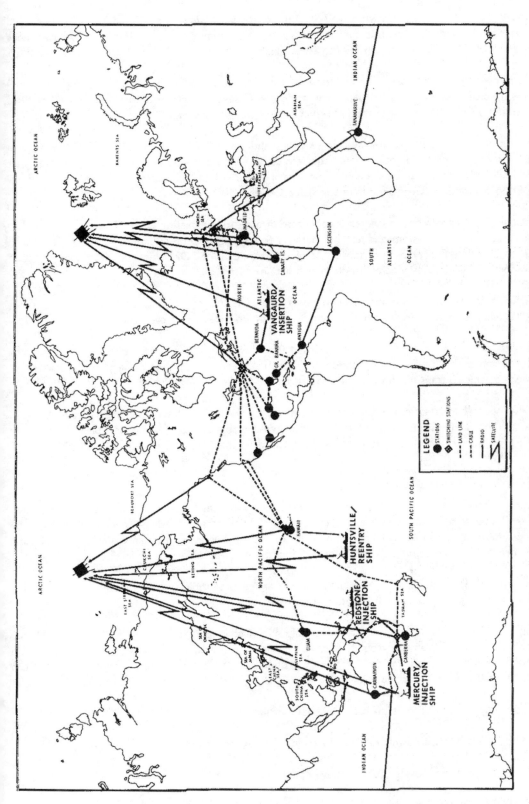

NASA COMMUNICATIONS NETWORK

NASA Communications Network

The NASA Communications Network (NASCOM) consists of several systems of diversely routed communications channels leased on communications satellites, common carrier systems and high frequency radio facilities where necessary to provide the access links.

The system consists of both narrow and wide-band channels, and some TV channels. Included are a variety of telegraph, voice, and data systems (digital and analog) with several digital data rates. Wide-band systems do not extend overseas. Alternate routes or redundancy provide added reliability.

A primary switching center and intermediate switching and control points provide centralized facility and technical control, and switching operations under direct NASA control. The primary switching center is at the Goddard Space Flight Center, Greenbelt, Md. intermediate switching centers are located at Canberra, Madrid, London, Honolulu, Guam, and Kennedy Space Center.

For Apollo 10, the Kennedy Space Center is connected directly to the Mission Control Center, Houston via the Apollo Launch Data System and to the Marshall Space Flight Center, Huntsville, Ala., by a Launch Information Exchange Facility. Both of these systems are part of NASCOM. They consist of data gathering and transmission facilities designed to handle launch data exclusively.

After launch, all network tracking and telemetry data hubs at GSFC for transmission to MCC Houston via two 50,000 bits-per second circuits used for redundancy and in case of data overflow.

Two Intelsat communications satellites will be used for Apollo 10. The Atlantic satellite will service the Ascension Island unified S-band (USB) station, the Atlantic Ocean ship and the Canary Islands site. These stations will be able to transmit through the satellite via the Comsat-operated ground station at Etam W.Va.

The second Apollo Intelsat communications satellite over the mid-Pacific will service the Carnarvon, Australia USB site and the Pacific Ocean ships. All these stations will be able to transmit simultaneously through the satellite to Houston via Brewster Flat, Wash., and the Goddard Space Flight Center, Greenbelt, Md.

Network Computers

At fraction-of-a-second intervals, the network's digital data processing systems, with NASA's Manned Spacecraft Center as the focal point, "talk" to each other or to the spacecraft. High-speed computers at the remote site (tracking ships included) issue commands or "up-link" data on such matters as control of cabin pressure, orbital guidance commands, or "go-no-go" indications to perform certain functions.

When information originates from Houston, the computers refer to their pre-programmed information for validity before transmitting the required data to the spacecraft.

Such "up-link" information is communicated by ultrahigh-frequency radio about 1,200 bit-per-second. Communication between remote ground sites, via high-speed communications links, occurs at about the same rate. Houston reads information from these ground sites at 2,400 bits-per-second, as well as from remote sites at 100 words-per-minute.

The computer systems perform many other functions, including:

Assuring the quality of the transmission lines by continually exercising data paths.

* Verifying accuracy of the messages by repetitive operations.

* Constantly updating the flight status.

For "down link" data, sensors built into the spacecraft continually sample cabin temperature, pressure, physical information on the astronauts such as heartbeat and respiration, among other items. These data are transmitted to the ground stations at 51.2 kilobits (12,800 binary digits) per-second.

At MCC the computers:

* Detect and select changes or deviations, compare with their stored programs, and indicate the problem areas or pertinent data to the flight controllers.

* Provide displays to mission personnel.

* Assemble output data in proper formats.

* Log data on magnetic tape for replay for the flight controllers.

* Keep time.

The Apollo Ships

The mission will be supported by four Apollo instrumentation ships operating as integral stations of the Manned Space Flight Network (MSFN) to provide coverage in areas beyond the range of land stations.

The ships, Vanguard, Redstone, Mercury, and Huntsville, will perform tracking, telemetry, and communication functions for the launch phase, Earth orbit insertion, translunar injection and reentry at the end of the mission.

Vanguard will be stationed about 1,000 miles southeast of Bermuda (25 degrees N, 49 degrees W) to bridge the Bermuda-Antigua gap during Earth orbit insertion. Vanguard also functions as part of the Atlantic recovery fleet in the event of a launch phase contingency. The Redstone, at 14 degrees S, 145.5 degrees E; Mercury, 32 degrees S, 131 degrees E; and Huntsville, 17 degrees S. 174 degrees W, provide a triangle of mobile stations between the MSFN stations at Carnarvon and Hawaii for coverage of the burn interval for translunar injection. In the event the launch data slips from May 18, the ships, will all move generally north eastward to cover the changing flight window patterns.

Redstone and Huntsville will be repositioned along the reentry corridor for tracking, telemetry, and communications functions during reentry and landing. They will track Apollo from about 1,000 miles away through communications blackout when the spacecraft will drop below the horizon and will be picked up by the ARIA aircraft.

The Apollo ships were developed jointly by NASA and the Department of Defense. The DOD operates the ships in support of Apollo and other NASA and DOD missions on a non-interference basis with Apollo requirements.

Management of the Apollo ships is the responsibility of the Commander, Air Force Western Test Range (AFWTR). The Military Sea Transport Service provides the maritime crews and the Federal Electric Corp., International Telephone and Telegraph, under contract to AFWTR, provides the technical instrumentation crews.

The technical crews operate in accordance with joint NASA/DOD standards and specifications which are compatible with MSFN operational procedures.

Apollo Range Instrumentation Aircraft (ARIA)

The Apollo Range Instrumentation Aircraft will support the mission by filling gaps in both land and ship station coverage where important and significant coverage requirements exist.

During Apollo 10, the ARIA will be used primarily to fill coverage gaps of the land and ship stations in the Indian Ocean and in the Pacific between Australia and Hawaii during the translunar injection interval. Prior to and during the burn, the ARIA record telemetry data from Apollo provide a real-time voice communication between the astronauts and the flight director at Houston.

Eight aircraft will participate in this mission, operating from Pacific, Australian and Indian Ocean air fields in positions under the orbital track of the spacecraft and booster. The aircraft, like the tracking ships, will be redeployed in a northeastward direction in the event of launch day slips.

For reentry, the ARIA will be redeployed to the landing area to continue communications between Apollo and Mission Control and provide position information on the spacecraft after the blackout phase of reentry has passed.

The total ARIA fleet for Apollo missions consist of eight EC-135A (Boeing 707) jets equipped specifically to meet mission needs. Seven-foot parabolic antennas have been installed in the nose section of the planes giving them a large, bulbous look.

The aircraft, as well as flight and instrumentation crews, are provided by the Air Force and they are equipped through joint Air Force-NASA contract action. ARIA operate in Apollo missions in accordance with MSFN procedures.

Ship Positions for Apollo 10

May 18, 1969

Insertion Ship (VAN)	25 degrees N - 49 degrees W
Injection Ship (MER)	32 degrees S - 131 degrees E
Injection Ship (RED)	14 degrees S - 145.5 degrees E
Injection Ship (RED)	20 degrees S - 172.5 degrees E
Reentry Support	
Reentry Ship (HTV)	17 degrees S - 174 degrees W

May 20, 1969

Insertion Ship (VAN)	25 degrees N - 49 degrees W
Injection Ship (MER)	32 degrees S - 131 degrees E
Injection Ship (RED)	14 degrees S - 145.5 degrees E
Injection ship (REP)	13 degrees S - 174 degrees E
Reentry Support	
Reentry Ship (HTV)	8 degrees S - 173 degrees W

May 23, 1969

Insertion Ship (VAN)	25 degrees N - 49 degrees W
Injection Ship (MER)	Released
Injection Ship (RED)	7.5 degrees S - 156 degrees E
Injection Ship (RED)	1 degree N - 177.5 degrees E
Reentry Support	
Reentry Ship (HTV)	10 degrees N- 172 degrees W

May 24, 1969

Insertion Ship (VAN)	25 degrees N - 49 degrees W
Injection Ship (MER)	Released
Injection Ship (RED)	3 degrees S - 158 degrees E
Injection Ship (RED)	9 degrees N - 175.5 degrees E
Reentry Support	
Reentry Ship (HTV)	15.5 degrees N - 173 degrees W

May 25, 1969

Insertion Ship (VAN)	25 degrees N - 49 degrees W
Injection Ship (MER)	Released
Injection Ship (RED)	0.5 degrees N - 161 degrees E
Injection Ship (RED)	16 degrees N - 174 degrees E
Reentry Support	
Reentry Ship (HTV)	22 degrees N - 173 degrees W

APOLLO PROGRAM MANAGEMENT

The Apollo Program, the United States' effort to land men on the Moon and return them safely to Earth before 1970, is the responsibility of the Office of Manned Space Flight (OMSF), National Aeronautics and Space Administration, Washington, D.C. Dr. George E. Mueller is Associate Administrator for Manned Space Flight.

NASA Manned Spacecraft Center (MSC), Houston, is responsible for development of the Apollo spacecraft, flight crew training and flight control. Dr. Robert R. Gilruth is Center Director.

NASA Marshall Space Flight Center (MSFC), Huntsville, Ala. is responsible for development of the Saturn launch vehicles. Dr. Wernher von Braun is Center Director.

NASA John F. Kennedy Space Center (KSC), Fla., is responsible for Apollo/Saturn launch operations. Dr. Kurt H. Debus is Center Director.

NASA Goddard Space Flight Center (GSFC), Greenbelt, Md., manages the Manned Space Flight Network under the direction of the NASA Office of Tracking and Data Acquisition (OTDA). Gerald M. Truszynski is Associate Administrator for Tracking and Data Acquisition. Dr. John F. Clark is Director of GSFC.

Apollo/Saturn Officials

NASA HEADQUARTERS

Lt. Gen. Sam C. Phillips, (USAF)	Apollo Program Director, OMSF
George H. Hage	Apollo Program Deputy Director, Mission Director, OMSF
Chester M. Lee	Assistant Mission Director, OMSF
Col. Thomas H. McMullen (USAF)	Assistant Mission Director, OMSF
Maj. Gen. James W. Humphreys, Jr.	Director of Space Medicine, OMSF
Norman Pozinsky	Director, Network Support Implementation Div., OTDA

Manned Spacecraft Center

George M. Low	Manager, Apollo Spacecraft Program
Kenneth S. Kleinknecht	Manager, Command and Service Modules
Brig. Gen. C. H. Bolender (USAF)	Manager, Lunar Module

Donald K. Slayton	Director of Flight Crew Operations
Christopher C. Kraft, Jr.	Director of Flight Operations
Glynn S. Lunney	Flight Director
Milton L. Windler	Flight Director
M. P. Frank	Flight Director
Gerald Griffin	Flight Director
Charles A. Berry	Director of Medical Research and Operations

Marshall Space Flight Center

Maj. Gen. Edmund F. O'Connor	Director of Industrial Operations
Dr. F.A. Speer	Director of Mission Operations
Lee B. James	Manager, Saturn V Program Office
William D. Brown	Manager, Engine Program Office

Kennedy Space Center

Miles Rose	Deputy Director, Center Operations
Rocco A. Petrone	Director, Launch Operations
Raymond L. Clark	Director, Technical Support
Rear Adm. Roderick O. Middleton (USN)	Manager, Apollo Program Office
Walter J. Kapryan	Deputy Director, Launch Operations
Dr. Hans F. Gruene	Director, Launch Vehicle Operations
John J. Williams	Directors Spacecraft Operations
Paul C. Donnelly	Launch Operations Manager

Goddard Space Flight Center

Ozro M. Covington	Assistant Director for Manned Space Flight Tracking
Henry F. Thompson	Deputy Assistant Director for Manned Space Flight Support
H. William Wood	Chief, Manned Flight Operations Div.
Tecwyn Roberts	Chief. Manned Flight Engineering Div.

Department of Defense

| Maj. Gen. Vincent G. Huston, (USAF) | DOD Manager of Manned Space |

Flight Support Operations

Maj. Gen. David M. Jones, (USAF)	Deputy DOD Manager of Manned Space Flight Support Operations, Commander of USAF Eastern Test Range
Rear Adm. F. E. Bakutis, (USK)	Commander of Combined Task Force 130, Pacific Recovery Area
Rear Adm. P. S. McManus, (USK)	Commander of Combined Task Force 140, Atlantic Recovery Area
Col. Royce G. Olson, (USAF)	Director of DOD Manned Space Flight Office
Brig. Gen. Allison C. Brooks, (USAF)	Commander Aerospace Rescue and Recovery Service

Major Apollo/Saturn V Contractors

Contractor	Item
Bellcomm Washington, D.C.	Apollo Systems Engineering
The Boeing Co. Washington, D.C.	Technical Integration and Evaluation
General Electric-Apollo Support Dept., Daytona Beach, Fla.	Apollo Checkout, and Quality and Reliability
North American Rockwell Corp. Space Div., Downey, Calif.	Command and Service Modules
Grumman Aircraft Engineering Corp., Bethpage, N.Y.	Lunar Module
Massachusetts Institute of Technology, Cambridge, Mass.	Guidance & Navigation (Technical Management)
General Motors Corp., AC Electronics Div., Milwaukee, Wis.	Guidance & Navigation (Manufacturing)
TRW Systems Inc. Redondo Beach, Calif.	Trajectory Analysis
Avco Corp., Space Systems	Heat Shield Ablative Material Div., Lowell, Mass.
North American Rockwell Corp. Rocketdyne Div. Canoga Park, Calif.	J-2 Engines., F-1 Engines
The Boeing Co. New Orleans	First Stage (SIC) of Saturn V, Launch Vehicles Saturn V Systems Engineering and Integration, Ground Support Equipment
North American Rockwell Corp. Space Div. Seal Beach, Calif.	Development and Production of Saturn V Second Stage (S-II)
McDonnell Douglas Astronautics Co. Huntington Beach, Calif.	Development and Production of Saturn V. Third Stage (S-IVB)
International Business Machines Federal Systems Div. Huntsville, Ala.	Instrument Unit
Bendix Corp. Navigation and Control Div. Teterboro, N.J.	Guidance Components for Instrument Unit (Including ST-124M Stabilized Platform)
Federal Electric Corp.	Communications and Instrumentation Support, KSC
Bendix Field Engineering Corp.	Launch Operations/complex Support, KSC
Catalytic-Dow	Facilities Engineering and Modifications, KSC

Hamilton Standard Division	Portable Life Support System;
United Aircraft Corp.	LM ECS
Windsor Looks, Conn. ILC Industries Dover, Del.	Space Suits
Radio Corp. of America Van Nuys, Calif.	110A Computer - Saturn Checkout
Sanders Associates Nashua, N.H.	Operational Display Systems Saturn
Brown Engineering Huntsville, Ala.	Discrete Controls
Reynolds, Smith and Hill Jacksonville., Fla.	Engineering Design of Mobile Launchers
Ingalls Iron Works Birmingham, Ala.	Mobile Launchers (ML) (structural work)
Smith/Ernst (Joint Venture) Tampa, Fla. /Washington, D. C.	Electrical Mechanical Portion of MLs
Power Shovel, Inc. Marion, Ohio	Transporter
Hayes International Birmingham, Ala.	Mobile Launcher Service Arms

APOLLO GLOSSARY

Ablating Materials — Special heat-dissipating materials on the surface of a spacecraft that vaporize during reentry.

Abort — The unscheduled termination of a mission prior to its completion.

Accelerometer — An instrument to sense accelerative forces and convert them into corresponding electrical quantities usually for controlling, measuring, indicating or recording purposes.

Adapter Skirt — A flange or extension of a stage or section that provides a ready means of fitting another stage or section to it.

Antipode — Point on surface of planet exactly 180 degrees opposite from reciprocal point on a line projected through center of body. In Apollo usage, antipode refers to a line from the center of the Moon through the center of the Earth and projected to the Earth surface on the opposite side. The antipode crosses the mid-Pacific recovery line along the 165th meridian of longitude once each 24 hours.

Apocynthion — Point at which object in lunar orbit is farthest from the lunar surface — object having been launched from body other than Moon. (Cynthia, Roman goddess of Moon)

Apogee — The point at which a Moon or artificial satellite in its orbit is farthest from Earth.

Apolune — Point at which object launched from the Moon into lunar orbit is farthest from lunar surfaces e.g.: ascent stage of lunar module after staging into lunar orbit following lunar landing.

Attitude — The position of an aerospace vehicle as determined by the inclination of its axes to some frame of reference; for Apollo, an inertial, space-fixed reference is used.

Burnout — The point when combustion ceases in a rocket engine.

Canard — A short, stubby wing-like element affixed to the launch escape tower to provide CM blunt end forward aerodynamic capture during an abort.

Celestial Guidance — The guidance of a vehicle by reference to celestial bodies.

Celestial Mechanics — The science that deals primarily with the effect of force as an agent in determining the orbital paths of celestial bodies.

Cislunar — Adjective referring to space between Earth and the Moon, or between Earth and Moon's orbit.

Closed Loop — Automatic control units linked together with a process to form an endless chain.

Deboost — A retrograde maneuver which lowers either perigee or apogee of an orbiting spacecraft. Not to be confused with deorbit.

Declination — Angular measurement of a body above or below celestial equator, measured north or south along the body's hour circle. Corresponds to Earth surface latitude.

Delta V — Velocity change.

Digital Computer — A computer in which quantities are represented numerically and which can be used to solve complex problems.

Down-Link — The part of a communication system that receives, processes and displays data from a spacecraft.

Entry Corridor — The final flight path of the spacecraft before and during Earth reentry.

Ephemeris — Orbital measurements (apogee, perigee, inclination, period, etc.) of one celestial body in relation to another at given times. In spaceflight, the orbital measurements of a spacecraft relative to the celestial body about which it orbited.

Escape Velocity — The speed a body must attain to overcome a gravitational field, such as that of Earth; the velocity of escape at the Earth's surface is 36,700 feet-per-second.

Explosive Bolts — Bolts destroyed or severed by a surrounding explosive charge which can be activated by an electrical impulse.

Fairing — A piece, part or structure having a smooth, streamlined outline, used to cover a non-streamlined object or to smooth a junction.

Flight Control System — A system that serves to maintain attitude stability and control during flight.

Fuel Cell — An electrochemical generator in which the chemical energy from the reaction of oxygen and a fuel is converted directly into electricity.

g or g Force — Force exerted upon an object by gravity or by reaction to acceleration or deceleration, as in a change of direction: one g is the measure of force required to accelerate a body at the rate of 32.16 feet-per-second.

Gimbaled Motor— A rocket motor mounted on gimbal; i.e.: on a contrivance having two mutually perpendicular axes of rotation, so as to obtain pitching and yawing correction moments.

Guidance System— A system which measures and evaluates flight information, correlates this with target data, converts the result into the conditions necessary to achieve the desired flight path, and communicates this data in the form of commands to the flight control system.

Heliocentric— Sun-centered orbit or other activity which has the Sun at its center.

Inertial Guidance— Guidance by means of the measurement and integration of acceleration from on board the spacecraft. A sophisticated automatic navigation system using gyroscopic devices, accelerometers etc., for high-speed vehicles. It absorbs and interprets such data as speed, position, etc., and automatically adjusts the vehicle to a pre-determined flight path. Essentially, it knows where it's going and where it is by knowing where it came from and how it got there. It does not give out any radio frequency signal so it cannot be detected by radar or jammed.

Injection— The process of boosting a spacecraft into a calculated trajectory.

Insertion— The Process of boosting a spacecraft into an orbit around the Earth or other celestial bodies.

Multiplexing— The simultaneous transmission of two or more signals within a single channel. The three basic methods of multiplexing involve the separation of signals by time division, frequency division and phase division.

Optical Navigation— Navigation by sight, as opposed to inertial methods, using stars or other visible objects an reference.

Oxidizer— In a rocket propellant, a substance such as liquid oxygen or nitrogen tetroxide which supports combustion of the fuel.

Penumbra— Semi-dark portion of a shadow in which light is partly cut off, e.g.: surface of Moon or Earth away from Sun where the disc of the Sun is only partly obscured.

Pericynthion— Point nearest Moon of object in lunar orbit— object having been launched from body other than Moon.

Perigee— Point at which a Moon or an artificial satellite in its orbit is closest to the Earth.

Perilune —The point at which a satellite (e.g.: a spacecraft) in its orbit is closest to the Moon. Differs from pericynthion in that the orbit is Moon-originated.

Pitch— The movement of a space vehicle about an axis (Y) that is perpendicular to its longitudinal axis.

Reentry— The return of a spacecraft that reenters the atmosphere after flight above it.

Retrorocket— A rocket that gives thrust in a direction opposite to the direction of the object's motion.

Right Ascension— Angular measurement of a body eastward along the celestial equator from the vernal equinox (0 degrees RA) to the hour circle of the body. Corresponds roughly to Earth surface longitude, except as expressed in hrs:min:sec instead of 180 degrees west and east from 0 degrees (24 hours=360 degrees).

Roll— The movements of a space vehicle about its longitudinal (X) axis.

S-Band— A radio-frequency band of 1,550 to 5,200 megahertz.

Selenographic— Adjective relating to physical geography of Moon. Specifically, positions on lunar surface as measured in latitude from lunar equator and in longitude from a reference lunar meridian.

Selenocentric— Adjective referring to orbit having Moon as center. (Selene, Gr. Moon)

Sidereal— Adjective relating to measurement of time, position or angle in relation to the celestial sphere and the vernal equinox.

State vector— Ground-generated spacecraft position, velocity and timing information uplinked to the spacecraft computer for crew use as a navigational reference.

Telemetering— A system for taking measurements within an aerospace vehicle in flight and transmitting them by radio to a ground station.

Terminator— Separation line between lighted and dark portions of celestial body which is not self luminous.

Ullage— The volume in a closed tank or container that is not occupied by the stored liquid; the ratio of this volume to the total volume of the tank; also an acceleration to force propellants into the engine pump intake lines before ignition.

Umbra— Darkest part of a shadow in which light is completely absent, e.g.: surface of Moon or Earth away from Sun where the disc of the Sun is completely obscured.

Update pad— Information on spacecraft attitudes, thrust values, event times, navigational data, etc., voiced up to the crew in standard formats according to the purpose, e.g.: maneuver update, navigation check, landmark tracking, entry update, etc.

Up-Link Data— Information fed by radio signal from the ground to a spacecraft.

Yaw— Angular displacement of a space vehicle about its vertical (Z) axis.

APOLLO ACRONYMS AND ABBREVIATIONS

(Note: This list makes no attempt to include all Apollo program acronyms and abbreviations, but several are listed that will be encountered frequently in the Apollo 10 mission. Where pronounced as words in air-to-ground transmissions, acronyms are phonetically shown in parentheses. Otherwise, abbreviations are sounded out by letter.)

AGS	(Aggs)	Abort Guidance System (LM)
AK		Apogee kick
APS	(Apps)	Ascent Propulsion System (LM)
		Auxiliary Propulsion System (S-IVB stage)
BMAG	(Bee-mag)	Body mounted attitude gyro
CDH		Constant delta height
CMC		Command Module Computer
COI		Contingency orbit insertion
CRS		Concentric rendezvous sequence
CSI		Concentric sequence initiate
DAP	(Dapp)	Digital autopilot
DEDA	(Dee-da)	Data Entry and Display Assembly (LM AGS)
DFI		Development flight instrumentation
DOI		Descent orbit insertion
DPS	(Dips)	Descent Propulsion system
DSKY	(Diskey)	Display and keyboard
EPO		Earth Parking Orbit

FDAI		Flight director attitude indicator
FITH	(Fith)	Fire in the hole (LM ascent abort staging)
FTP		Fixed throttle point
HGA		High-gain antenna
IMU		Inertial measurement unit
IRIG	(Ear-ig)	Inertial rate integrating gyro
LOI		Lunar orbit insertion
LPO		Lunar parking orbit
MCC		Mission Control Center
MC&W		Master caution and warning
MSI		Moon sphere of influence
MTVC		Manual thrust vector control
NCC		Combined corrective maneuver
NSR		Coelliptical maneuver
PIPA	(Pippa)	Pulse integrating pendulous accelerometer
PLSS	(Pliss)	Portable life support system
PTC		Passive thermal control
PUGS	(pugs)	Propellant utilization and gauging system
REFSMMAT	(Refsmat)	Reference to stable member matrix
RHC		Rotation hand controller
RTC		Real-time command
SCS		Stabilization and control system
SLA	(Slah)	Spacecraft LM adapter
SPS		Service propulsion system
TEI		Transearth injection
THC		Thrust hand controller
TLI		Translunar injection
TPF		Terminal phase finalization
TPI		Terminal phase initiate
TVC		Thrust vector control

CONVERSION FACTORS

Multiply	By	To Obtain

Distance:

feet	0.3048	meters
meters	3.281	feet
kilometers	3281	feet
kilometers	0.6214	statute miles
statute miles	1.609	kilometers
nautical miles	1.852	kilometers
nautical miles	1.1508	statute miles
statute miles	0.86898	nautical miles
statute mile	1760	yards

Velocity:

feet/sec	0.3048	meters/sec
meters/sec	3.281	feet/sec
meters/sec	2.237	statute mph
feet/sec	0.6818	statute miles/hr
feet/sec	0.5925	nautical miles/hr
statute miles/hr	1.609	km/hr

| nautical miles/hr (knots) | 1.852 | km/hr |
| km/hr | 0.6214 | statute miles/hr |

Liquid measure, weight:

gallons	3.785	liters
liters	0.2642	gallons
pounds	0.4536	kilograms
kilograms	2.205	pounds

Volume:

| cubic feet | 0.02832 | cubic meters |

Pressure:

| pounds/sq inch | 70.31 | grams/sq cm |

Commander Tom Stafford poses with the Lunar Module's namesake "Snoopy".

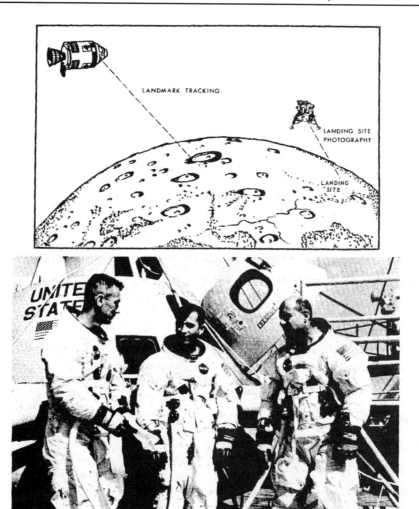

LANDMARK TRACKING

LANDING SITE PHOTOGRAPHY

LANDING SITE

LM PILOT
E. A. CERNAN

CM PILOT
J. W. YOUNG

COMMANDER
T. P. STAFFORD

MISSION OPERATION REPORT
APOLLO 10 (AS-505) MISSION

OFFICE OF MANNED SPACE FLIGHT
Prepared by: Apollo Program Office-MAO
FOR INTERNAL USE ONLY

Pre-Launch
Mission Operation Report
No. M-932-69-10

MEMORANDUM

To: A/Administrator

From: MA/Apollo Program Director

Subject: Apollo 10 Mission (AS-505)

No earlier than 18 May 1969, we plan to launch the next Apollo/Saturn V mission, Apollo 10. This will be the third manned Saturn V flight, the fourth flight of a manned Apollo Command/Service Module, and the second flight of a manned Lunar Module.

The Apollo 10 Mission will be a manned lunar mission development flight. It will demonstrate crew/ space vehicle/ mission support facilities performance during a manned lunar mission with the Command/Service Module and Lunar Module, and will evaluate Lunar Module performance in the cislunar and lunar environment. The mission will be about eight days in duration and will use the operational configuration of the Saturn V Launch Vehicle and Apollo Spacecraft.

Apollo 10 will be the first Saturn V flight launched from Pad B of Launch Complex 39 at the Kennedy Space Center. Five daily launch windows per month have been selected for May, June, and July. The first three days of each monthly window provide favorable lunar lighting conditions for the primary Apollo 11 Mission landing sites. The last two days of each monthly window provide slightly degraded lunar lighting conditions. Lunar operations will simulate the Apollo 11 Mission timeline as closely as possible. Recovery will be in the Pacific Ocean at 165 degrees west longitude and at southern latitudes for the first two days of each monthly window and at 175 degrees west longitude and northern latitudes thereafter.

Sam C. Phillips
Lt. General, USAF
Apollo Program Director

Associate Administrator for Manned Space Flight

FOREWORD

MISSION OPERATION REPORTS are published expressly for the use of NASA Senior Management, as required by the Administrator in NASA Instruction 6-2-10, dated August 15, 1963. The purpose of these reports is to provide NASA Senior Management with timely, complete, and definitive information on flight mission plans, and to establish official mission objectives which provide the basis for assessment of mission accomplishment.

Initial reports are prepared and issued for each flight project just prior to launch. Following launch, updating reports for each mission are issued to keep General Management currently informed of definitive mission results as provided in NASA Instruction 6-2-10.

Because of their sometimes highly technical orientation, distribution of these reports is limited to personnel having program-project management responsibilities. The Office of Public Affairs publishes a comprehensive series of pre-launch and post launch reports on NASA flight missions, which are available for general distribution.

The Apollo 10 Mission Operation Report is published in two volumes: the Mission Operation Report (MOR); and the Mission Operation Report Supplement. This format was designed to provide a mission-oriented document in the MOR with only a very brief description of the space vehicle and support facilities. The MOR Supplement is a program-oriented reference document with a more comprehensive description of the space vehicle, launch complex, and mission monitoring, support, and control facilities. The MOR Supplement was issued on 25 February 1969 for the previous mission, Apollo 9, and will not be reissued for this mission.

Published and Distributed by PROGRAM and SPECIAL REPORTS DIVISION (XP) EXECUTIVE SECRETARIAT - NASA HEADQUARTERS

THE APOLLO 10 MISSION

The Apollo 10 Mission will be a manned lunar mission development flight. It will demonstrate crew/space vehicle/mission support facilities performance during a manned lunar mission with the Command/Service Module (CSM) and Lunar Module (LM), and will evaluate LM performance in the cislunar and lunar environment. In addition, more knowledge of the lunar gravitational effect, additional refinement of Manned Space Flight Network tracking techniques, and landmark tracking will be obtained. The mission will be about eight days in duration and will use the operational configuration of the Saturn V Launch Vehicle and Apollo Spacecraft with the LM Ascent Propulsion System (APS) propellant tanks loaded to half capacity.

The mission profile through descent orbit insertion will be similar to the first lunar landing mission including targeting for specific lunar landing sites. During the LM active phase, the manned LM will perform a minimum energy descent to approximately 50,000 feet above the lunar surface. The LM will then perform a phasing revolution to make the required adjustment in the CSM lead angle prior to orbit insertion. The rendezvous maneuvers will be performed as on a lunar landing mission. Prior to phasing, the lunar surface will be photographed from the LM. After the LM-active phase, an unmanned LM APS burn to propellant depletion will be performed. Between the APS depletion burn and the transearth injection burn, lunar surface photography and lunar landmark tracking from the CSM will be accomplished.

Earth touchdown will be in the Pacific Ocean at 165°W longitude within +35° latitude and will nominally occur eight days from launch. The recovery line shifts-westward starting at 00 latitude until it reaches 175°W longitude at 15°N latitude and then continues north to 35°N latitude.

The Apollo 10 Mission will provide the following first-time in-flight opportunities:

Lunar orbit rendezvous.

Docked lunar landmark tracking.

LM steerable antenna operation at distances greater than those of low earth orbit enabling its evaluation under conditions for which it was designed.

Descent Propulsion System engine burn in the lunar landing mission configuration and environment.

Lunar landing mission profile simulation (except for powered descent, lunar surface activity, and ascent).

Low level (50,000 feet) evaluation of lunar visibility.

Docked CSM/LM thermal control in the absence of earth albedo and during long periods of sunlight.

LM omni-directional antenna operation at lunar distance.

Abort Guidance System operation during an APS burn over the range of inertias for a lunar mission.

VHF ranging during a rendezvous.

Landing radar operation near lunar environment where the reflected energy from the lunar surface will be detected.

Transposition, docking, and LM ejection in daylight after the S-IVB burn when the S-IVB is in inertial hold attitude and while the spacecraft is moving away from the earth.

Translunar midcourse correction with a docked CSM/LM.

LM Digital Uplink Assembly first flight (replaces Digital Command Assembly used on LM-3).

PROGRAM DEVELOPMENT

The first Saturn vehicle was successfully flown on 27 October 1961 to initiate operations of the Saturn I Program. A total of 10 Saturn I vehicles (SA-1 to SA-10) was successfully flight tested to provide information on the integration of launch vehicle and spacecraft and to provide operational experience with large multi-engined booster stages (S-1, S-IV).

The next generation of vehicles, developed under the Saturn IB Program, featured an uprated first stage (S-1B) and a more powerful new second stage (S-IVB). The first Saturn IB was launched on 26 February 1966. The first three Saturn IB missions (AS-201, AS-203, and AS-202) successfully tested the performance of the launch vehicle and spacecraft combination, separation of the stages, behavior of liquid hydrogen in a weightless environment, performance of the Command Module heat shield at low earth orbital entry conditions, and recovery operations.

The planned fourth Saturn IB mission (AS-204) scheduled for early 1967 was intended to be the first manned Apollo flight. This mission was not flown because of a spacecraft fire, during a manned prelaunch test, that took the lives of the prime flight crew and severely damaged the spacecraft. The SA-204 Launch Vehicle was later assigned to the Apollo 5 Mission.

The Apollo 4 Mission was successfully executed on 9 November 1967. This mission initiated the use of the Saturn V Launch Vehicle (SA-501) and required an orbital restart of the S-IVB third stage. The spacecraft for this mission consisted of an unmanned Command/Service Module (CSM) and a Lunar Module test article (LTA). The CSM Service Propulsion System (SPS) was exercised, including restart, and the Command Module Block 11 heat shield was subjected to the combination of high heat load, high heat rate, and aerodynamic loads representative of lunar return entry. All primary mission objectives were successfully accomplished.

The Apollo 5 Mission was successfully launched and completed on 22 January 1968. This was the fourth mission utilizing Saturn IB vehicles (SA-204). This flight provided for unmanned orbital testing of the Lunar Module (LM-1). The LM structure, staging, and proper operation of the Lunar Module Ascent Propulsion System (APS) and Descent Propulsion System (DPS), including restart, were verified. Satisfactory performance of the S-IVB/Instrument Unit (IU) in orbit was also demonstrated. All primary objectives were achieved.

The Apollo 6 Mission (second unmanned Saturn V) was successfully launched on 4 April 1968. Some flight anomalies encountered included oscillation reflecting propulsion structural longitudinal coupling, an imperfection in the Spacecraft LM Adapter (SLA) structural integrity, and certain malfunctions of the J-2 engines in the S-II and S-IVB stages. The spacecraft flew the planned trajectory, but preplanned high velocity reentry conditions were not achieved. A majority of the mission objectives for Apollo 6 were accomplished.

The Apollo 7 Mission (first manned Apollo) was successfully launched on 11 October 1968. This was the fifth and last planned Apollo mission utilizing a Saturn IB Launch Vehicle (SA-205). The 11-day mission provided the first orbital tests of the Block II Command/Service Module. All primary mission objectives were successfully accomplished. In addition, all planned detailed test objectives, plus three that were not originally scheduled, were satisfactorily accomplished.

The Apollo 8 Mission was successfully launched on 21 December and completed on 27 December 1968. This was the first manned flight of the Saturn V Launch Vehicle and the first manned flight to the vicinity of the moon. All primary mission objectives were successfully accomplished. In addition, all detailed test objectives plus four that were not originally scheduled, were successfully accomplished. Ten orbits of the moon were successfully performed with the last eight circular at an altitude of 60 nm. TV and photographic coverage was successfully carried out, with telecasts to the public being made in real time.

The Apollo 9 Mission was successfully launched on 3 March and completed on 13 March 1969. This was the second manned Saturn V flight, the third flight of a manned Apollo Command/Service Module, and the first flight of a manned Lunar Module. This flight provided the first manned LM systems performance demonstration. All primary mission objectives were successfully accomplished. All detailed test objectives were accomplished except two associated with S-band and VHF communications which were partially accomplished. The S-IVB second orbital restart, CSM transposition and docking, and LM rendezvous and docking were also successfully demonstrated.

NASA OMSF PRIMARY MISSION OBJECTIVES FOR APOLLO 10

PRIMARY OBJECTIVES

Demonstrate crew/space vehicle/mission support facilities performance during a manned lunar mission with CSM and LM.

Evaluate LM performance in the cislunar and lunar environment.

Sam C. Phillips
Lt. General, USAF
Apollo Program Director

Date: - 6 May 1969

George W. Mueller
Associate Administrator for
Manned Space Flight

Date: 8 May 1969

The crew of Apollo 10 (l to r) Eugene Cernan,
Thomas Stafford and John Young.

Lunar Module 4 in the Kennedy Space
Center's Manned Spacecraft Operations
Building being moved into position for
mating with Spacecraft Lunar Module
Adapter (SLA) 13. (below left)

Apollo 10 ready for launch. (below)

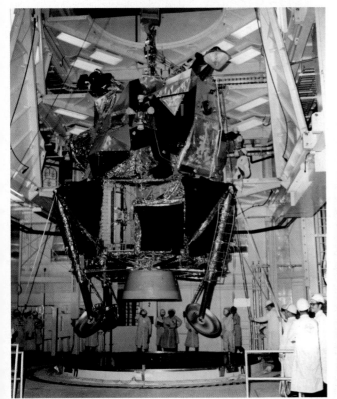

Young, Stafford and Cernan at the Cape during emergency egress training. (above)

Apollo 10 before and during launch. (below) The five F-1 engines were the largest and most powerful rockets ever built, generating an astonishing seven million pounds of thrust at lift-off and burning almost 15 tons of fuel per second for their expected life-span of just under three minutes.

One of the most experienced crews to fly, all three men on Apollo 10 had flown previously on the two-man Gemini programme. Cernan (left) had been one of the first men to walk in space and subsequently walked on the moon, while Young (center) not only walked on the moon but also became the only Apollo veteran to fly the Space Shuttle. Tom Stafford, a veteran of four space rendezvous' commanded the Lunar Module "Snoopy" to within eight miles of the lunar surface before firing the ascent engine and returning to the orbiting Command Module "Charlie Brown".

John Young and Eugene Cernan practice exiting into a life-raft from a mock-up Command Module prior to the flight.

A stupendous shot of the Earth showing most of North Africa.

The crew of Apollo 10 used up most of their 70mm film to take pictures of the moon but they also brought home an array of beautiful pictures of the Earth such as this one and the one on the following page which shows Baja California.

Few pictures were taken of the undocking and descent procedures in Lunar orbit. These artists renditions show the Lunar Module on it's way down to an altitude of eight miles using the hypergolic-fuel-driven descent engine.

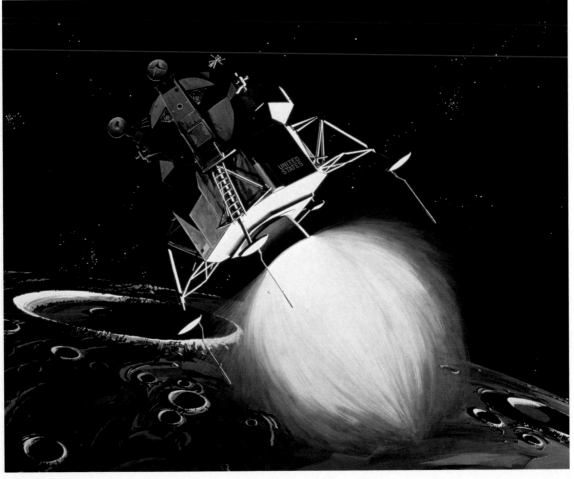

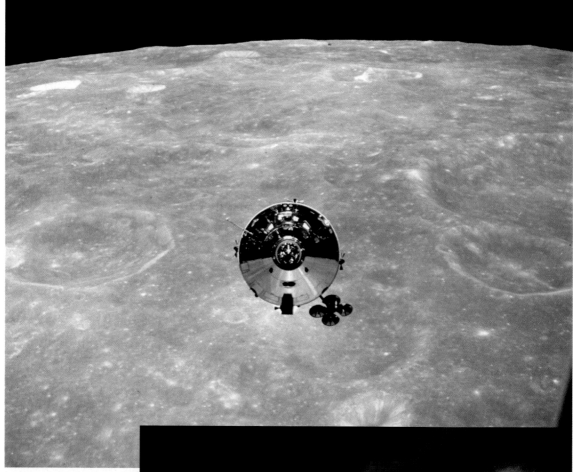

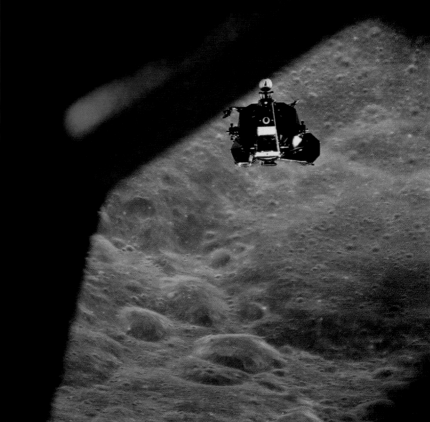

Stirring and dramatic pictures of two spacecraft in orbit around the moon.

Apollo 10 was the first flight which had an opportunity to show the Apollo hardware flying in Lunar orbit.

Lunar Module "Snoopy" is shown returning for docking after the jettisoning of the descent stage went awry due to the Abort Guidance System being left in AUTO mode. Stafford was required to switch to manual to complete the maneuver.

Retrieval of Apollo 10 was swift as the vehicle landed within three miles of the recovery vessel USS Princeton (right)

A US Navy recovery helicopter deploys frogmen into the water around the Command Module "Charlie Brown" shortly after splashdown. (left)

A shot taken from the above helicopter shows Young & Stafford aboard the rubber liferaft. One astronaut is reaching for the cage which is hanging from the hovering copter.

After a nearly flawless flight the crew of Apollo 10 arrive aboard the USS Princeton (left).

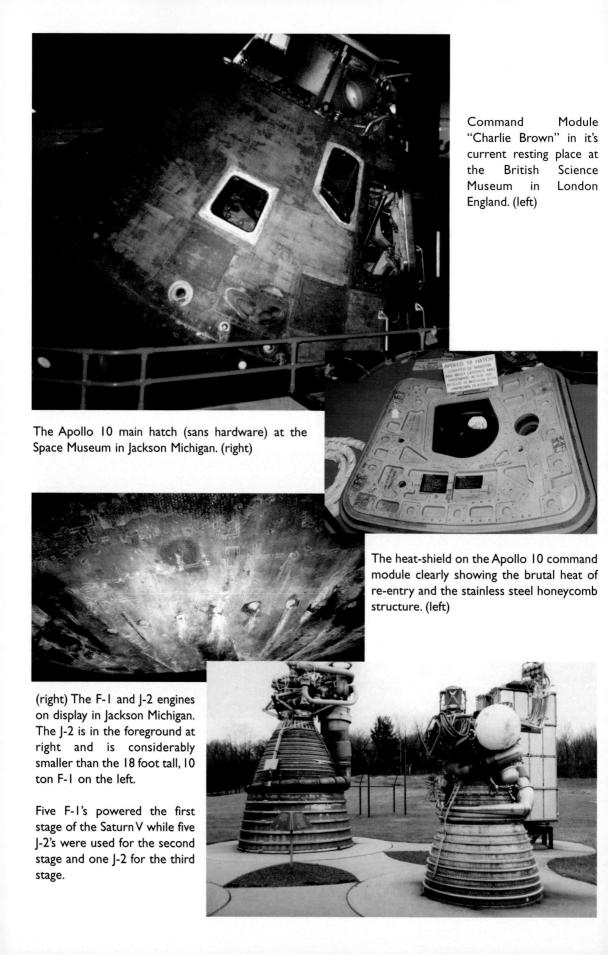

Command Module "Charlie Brown" in it's current resting place at the British Science Museum in London England. (left)

The Apollo 10 main hatch (sans hardware) at the Space Museum in Jackson Michigan. (right)

The heat-shield on the Apollo 10 command module clearly showing the brutal heat of re-entry and the stainless steel honeycomb structure. (left)

(right) The F-1 and J-2 engines on display in Jackson Michigan. The J-2 is in the foreground at right and is considerably smaller than the 18 foot tall, 10 ton F-1 on the left.

Five F-1's powered the first stage of the Saturn V while five J-2's were used for the second stage and one J-2 for the third stage.

DETAILED TEST OBJECTIVES

The Detailed Test Objectives (DTO's) listed below have been assigned to the Apollo 10 Mission. Principal DTO's are planned for accomplishment on the Apollo 10 Mission in order to demonstrate a lunar landing capability. Secondary DTO's are not prerequisites to a lunar landing mission, but will provide significant data or experience. No mandatory DTO's will be performed on this mission.

LAUNCH VEHICLE*

Secondary DTO's

Verify J-2 engine modifications.

Confirm J-2 engine environment in S-II and S-IVB stages.

Confirm launch vehicle longitudinal oscillation environment during S-IC stage burn period.

Verify that modifications incorporated in the S-IC stage suppress low frequency longitudinal oscillations.

Confirm launch vehicle longitudinal oscillation environment during S-II stage burn period.

Demonstrate that early center engine cutoff for S-II stage suppresses low frequency longitudinal oscillations.

No principal DTO's have been assigned.

SPACECRAFT

Principal DTO's

Demonstrate CSM/LM rendezvous capability for a lunar landing mission (20.78).

Perform manual and automatic acquisition, tracking, and communications with MSFN using the steerable S-band antenna at lunar distance (16. 10).

Perform lunar landmark tracking from the CSM while in lunar orbit (20.121).

Perform lunar landmark tracking in lunar orbit from the CSM with the LM attached (20.91).

Operate the landing radar at the closest approach to the moon and during DPS burns (16.14).

Obtain data on the CM and LM crew procedures and timeline for the lunar orbit phase of a lunar landing mission (20.66).

Perform PGNCS/DPS undocked Descent Orbit Insertion (DOI) and a high thrust maneuver (11.15).

Secondary DTO's

Demonstrate LM/CSM/MSFN communications at lunar distance (16.17).

Communicate with MSFN using the LM S-band omni-antennas at lunar distance (16.12).

Obtain data on the rendezvous radar performance and capability near maximum range (16.15).

Obtain supercritical helium system pressure data while in standby conditions and during all DPS engine firings (13.14).

Perform an unmanned AGS-controlled APS burn (12.9).

Obtain data on the operational capability of VHF ranging during a LM-active rendezvous (20.77).

Obtain data on the effects of lunar illumination and contrast conditions on crew visual perception while in lunar orbit (20.86).

Obtain data on the Passive Thermal Control (PTC) system during a lunar orbit mission (7.26).

Demonstrate CSM/LM passive thermal control modes during a lunar orbit mission (20.79).

Demonstrate RCS translation and attitude control of the staged LM using automatic and manual AGS/CES control (12.8).

Evaluate the ability of the AGS to perform a LM-active rendezvous (12.10).

Monitor PGNCS/AGS performance during lunar orbit operations (20.82).

Demonstrate operational support for a CSM/LM lunar orbit mission (20.80).

Perform a long duration unmanned APS burn (13.13).

Perform lunar orbit insertion using SPS GNCS-controlled burns with a docked CSM/LM (20.117).

Obtain data to verify IMU performance in the flight environment (11.17).

Perform a reflectivity test using the CSM S-band high-gain antenna while docked (6.9).

Perform CSM transposition, docking, and CSM/LM ejection after the S-IVB TLI burn (20.46).

Perform translunar midcourse corrections (20.95).

Obtain AGS performance data in the flight environment (12.6).

Perform star-lunar landmark sightings during the transearth phase (1.39).

Obtain data on LM consumables for a simulated lunar landing mission, in lunar orbit, to determine lunar landing mission consumables (20.83).

LAUNCH COUNTDOWN AND TURNAROUND CAPABILITY, AS-505

COUNTDOWN

Countdown for the Apollo 10 Mission will begin with a precount period starting at T-93 hours during which Launch Vehicle (LV) and Spacecraft (SC) countdown activities will be independently conducted. Coordinated SC and LV launch countdown will contain only a single built-in hold (6 hours at T-9). It is anticipated that additional holds (assuming slack time is available) will be inserted at advantageous times after the Count Down Demonstration Test has been completed. Figure 1 shows the significant launch countdown events.

SCRUB/TURNAROUND

The space vehicle turnaround will begin immediately following a scrub during the countdown. Turnaround is the time required to recycle and countdown to launch (T-0). A 3-day scrub turnaround is planned for AS-505. Activity times are considered to be minimal and do not account for serial time which may be required for repair and retest of any system which may have caused the scrub.

Six primary cases can be identified to implement the required turnaround activities in preparation for a subsequent launch attempt following a countdown scrub prior to ignition. These cases identify the turnaround activities necessary to maintain the same confidence for subsequent launch attempts as for the original attempt. The six cases are:

Case I
Scrub/turnaround at post-LV cryogenic loading - CSM/LM cryogenic reservicing (65 hours 45 minutes).

Case 2.
Scrub/turnaround at post-LV cryogenic loading - LM cryogenic reservicing (39 hours 15 minutes).

Case 3.
Scrub/turnaround at post-LV cryogenic loading - LM cryogenic reservicing (23 hours 15 minutes).

Case 4.
Scrub/turnaround at pre-LV cryogenic loading - CSM/LM cryogenic reservicing (55 hours 30 minutes).

Case 5.
Scrub/turnaround at pre-LV cryogenic loading - LM cryogenic reservicing (32 hours 00 minutes).

Case 6.
Scrub/turnaround at pre-LV cryogenic loading - No CSM/LM cryogenic reservicing (hold for next launch window).

A second scrub/turnaround would result in a combination of cases and each combination presents special considerations.

*LAUNCH COUNTDOWN, AS-505

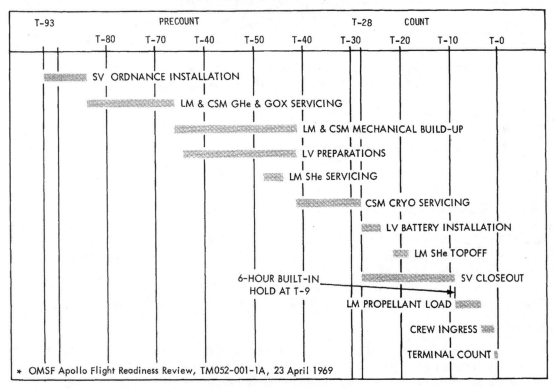

* OMSF Apollo Flight Readiness Review, TM052-001-1A, 23 April 1969

DETAILED FLIGHT MISSION DESCRIPTION

LAUNCH WINDOWS

Launch windows are based on range safety flight azimuth limits of 72° to 108°. No principal DTO's have been assigned (based on the earth-fixed heading of the launch vehicle at the beginning of the pitch program), on booster and spacecraft performance, on insertion tracking, and on meeting lighting constraints at the candidate lunar landing sites. The minimum operational daily window is 2½ hours. Mission planning allows for launch attempts during each of three consecutive months, starting at the planning date in the Apollo launch schedule. Current guidelines call for mission plans to be based on sites (and launch windows) approved for the initial lunar landing mission.

LUNAR LANDING SITES

The following landing sites have been approved for mission planning:

				MAY (EDT)	
SITE	LAT.	LONG.	DATE	OPEN-CLOSE	SEA
2	0°44'N	23°39'E	18	12:49-17:09	11°
3	0°22'N	1°21'W	20	13:03-17:24	10.5°
4	3°39'S	36°42'W	23	13:12-17:35	10°
5	1°46'N	41°56'W	24	13:15-17:40	17°
5	1°46'N	41°56'W	25	13:19-17:45	28°*

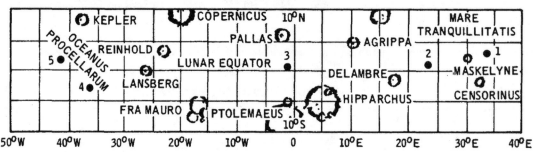

* HIGH SUN ELEVATION ANGLE (SEA) IS ACCEPTED AT SITE 5 TO EXTEND LAUNCH WINDOW TO FIVE OPPORTUNITIES.

LUNAR LANDING SITES

The following landing sites have been approved for mission planning:

NOMINAL MISSION

The launch date set for Apollo 10 is 18 May 1969 at 12:49 EDT. The launch azimuth will be 72 degrees, translunar injection will occur during the second orbit over the Pacific Ocean, and the targeted lunar landing site will be Site 2. The duration of this mission will be approximately 8 days with a lunar orbital stay time of about 61.5 hours. Transearth flight time will be approximately 53.5 hours and touchdown will occur in the Pacific Ocean at 165°W longitude, 15°S latitude.

A summary flight profile of the Apollo 10 Mission is shown in Figure 2 and the summary flight plan is shown in Figure 3.

The sequence of events for the Apollo 10 Mission is given in Table I. Launch Vehicle (LV) Time Base (TB) notations are also included. TB's may be defined as precise initial points upon which succeeding critical preprogrammed activities or functions may be based. The TB's noted in Table I are for a nominal mission and presuppose nominal LV performance. However, should the launch vehicle stages produce non-nominal performance, the launch vehicle computer will recompute the subsequent TB's and associated burns to correct LV performance to mission rules.

First Period of Activity

The Saturn V Launch Vehicle will place the following vehicle combination into a 103-nm circular earth parking orbit: S-IVB stage, Instrument Unit (IU), Lunar Module (LM), Spacecraft LM Adapter (SLA) and Command/Service Module (CSM). The ascent trajectory is shown in Figure 4. The launch time and azimuth are chosen to support two translunar injection (TLI) opportunities for an optimized payload. Checkout of the S-IVB, IU, and CSM will be accomplished during this orbital coast period. The earth orbital configuration is shown in Figure 5.

The S-IVB J-2 engine will be reignited during the second parking orbit to inject the vehicle combination into a translunar trajectory. Within 2 hours after injection, the CSM will be separated from the remainder of the vehicle and will transpose, dock with the LM, and initiate ejection of the CSM/LM from the SLA/IU/S-IVB as shown in Figure 6. A pitchdown maneuver of a prescribed magnitude for this transposition, docking, and ejection (TD&E) phase is designed to place the sun over the shoulders of the crew, avoiding CSM shadow on the docking interface. The pitch maneuver also provides continuous tracking and communications during the inertial attitude hold during TD&E. A Service Propulsion System (SPS) evasive maneuver will place the docked spacecraft, as shown in Figure 7, on a free-return trajectory. A free-return to earth will be possible if the insertion into lunar parking orbit cannot be accomplished. Land landing is dictated by the free-return constraints for some opportunities, but can be avoided by a corrective maneuver at an acceptable time during the translunar coast. Any necessary spacecraft midcourse corrections (MCC) will be accomplished during translunar coast. These corrections will utilize the Manned Space Flight Network (MSFN) for navigation.

APOLLO 10 FLIGHT PROFILE

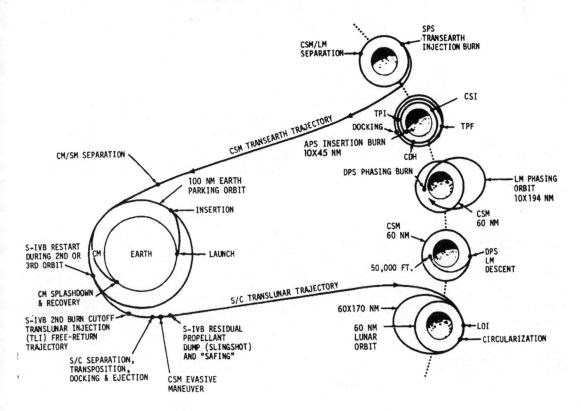

APOLL(
SUMMARY FL

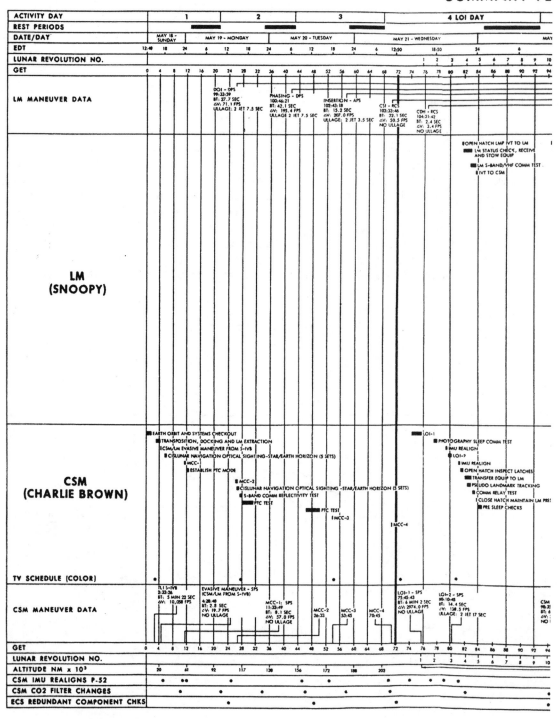

ACTIVITY DAY		1	2	3	4 LOI DAY	
REST PERIODS						
DATE/DAY		MAY 18 - SUNDAY	MAY 19 - MONDAY	MAY 20 - TUESDAY	MAY 21 - WEDNESDAY	MAY
EDT	12:48	18 24	6 12 18 24	6 12 18 24	6 12:50 18:50 24	6
LUNAR REVOLUTION NO.					1 2 3 4 5 6 7 8 9 10	
GET		0 4 8 12 16 20 24 28 32	36 40 44 48 52 56 60	64 68 72 74 76 78 80	82 84 86 88 90 92 94	

LM MANEUVER DATA
- DOI - DPS 99:33:59 BT: 27.7 SEC ΔV: 71.1 FPS ULLAGE: 2 JET 7.5 SEC
- PHASING - DPS 100:46:21 BT: 42.1 SEC ΔV: 195.4 FPS ULLAGE 2 JET 7.5 SEC
- INSERTION - APS 102:43:18 BT: 15.2 SEC ΔV: 207.0 FPS ULLAGE: 2 JET 3.5 SEC
- CSI - RCS 103:33:46 BT: 32.1 SEC ΔV: 50.5 FPS NO ULLAGE
- CDH - RCS 104:31:42 BT: 2.4 SEC ΔV: 3.4 FPS NO ULLAGE

LM (SNOOPY)
- OPEN HATCH LMP IVT TO LM
- LM STATUS CHECK, RECEIVE AND STOW EQUIP
- LM S-BAND/VHF COMM TEST
- IVT TO CSM

CSM (CHARLIE BROWN)
- EARTH ORBIT AND SYSTEMS CHECKOUT
- TRANSPOSITION, DOCKING AND LM EXTRACTION
- CSM/LM EVASIVE MANEUVER FROM S-IVB
- CISLUNAR NAVIGATION OPTICAL SIGHTING-STAR/EARTH HORIZON (5 SETS)
- MCC-1
- ESTABLISH PTC MODE
- MCC-2
- CISLUNAR NAVIGATION OPTICAL SIGHTING -STAR/EARTH HORIZON (5 SETS)
- S-BAND COMM REFLECTIVITY TEST
- PTC TEST
- PTC TEST
- MCC-3
- MCC-4
- LOI-1
- PHOTOGRAPHY SLEEP COMM TEST
- IMU REALIGN
- LOI-7
- IMU REALIGN
- OPEN HATCH INSPECT LATCHES
- TRANSFER EQUIP TO LM
- PSEUDO LANDMARK TRACKING
- COMM RELAY TEST
- CLOSE HATCH MAINTAIN LM PRES
- PRE SLEEP CHECKS

TV SCHEDULE (COLOR)	•	•	•	•

CSM MANEUVER DATA
- TLI S-IVB 2:33:26 BT: 5 MIN 22 SEC ΔV: 10,058 FPS
- EVASIVE MANEUVER - SPS (CSM/LM FROM S-IVB) 4:28:48 BT: 2.8 SEC ΔV: 19.7 FPS NO ULLAGE
- MCC-1: SPS 11:33:49 BT: 8.1 SEC ΔV: 57.0 FPS NO ULLAGE
- MCC-2 26:33
- MCC-3 53:45
- MCC-4 70:45
- LOI-1 - SPS 75:45:43 BT: 6 MIN 2 SEC ΔV: 2974.0 FPS NO ULLAGE
- LOI-2 - SPS 80:10:45 BT: 14.4 SEC ΔV: 138.5 FPS ULLAGE: 2 JET 17 SEC
- CSM 98:35 BT: 6 ΔV: NO

GET		0 4 8 12 16 20 24 28 32	36 40 44 48 52 56 60	64 68 72 74 76 78 80	82 84 86 88 90 92 94	
LUNAR REVOLUTION NO.					1 2 3 4 5 6 7 8 9 10	
ALTITUDE NM x 10³		20 61 92	117 138 156	172 188 202		
CSM IMU REALIGNS P-52		• ••	• •	•	• • • • • •	
CSM CO2 FILTER CHANGES		•	• •	• •	• •	•
ECS REDUNDANT COMPONENT CHKS		•	•	•	•	•

LLO 10
FLIGHT PLAN

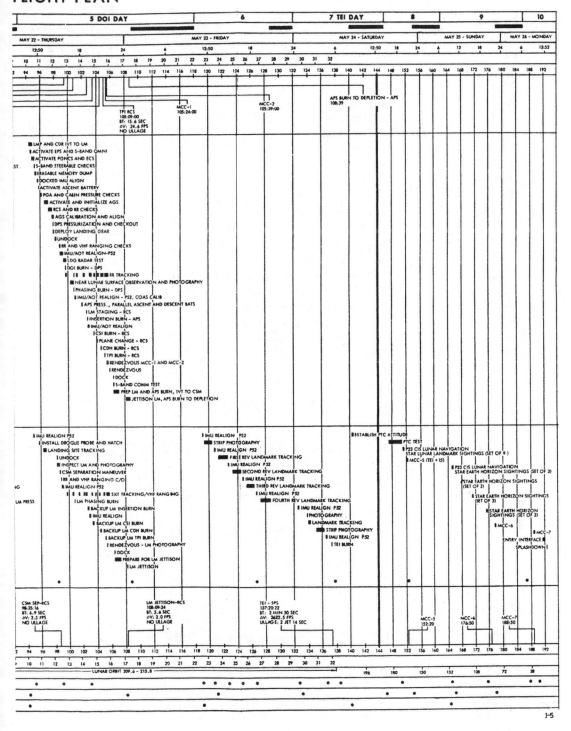

1-5

TABLE I

APOLLO 10 SEQUENCE OF EVENTS*

Ground Elapsed Time (GET) HR:MIN:SEC:	Event
00:00:00	Lift-off - Time Base (TB) - I
00:00:30	SV Roll Complete
00:01:17	Max Q (Maximum Dynamic Pressure)
00:02:15	S-IC Inboard Engine Cutoff - TB2
00:02:40	S-IC Outboard Engine Cutoff - TB3
00:02:42	S-IC/S-II Separation
00:02:42	S-II Ignition
00:03:16	Launch Escape Tower Jettison
00:09:14	S-II Engine Cutoff - TB4
00:09:15	S-II/S-IVB Separation
00:09:18	S-IVB Ignition
00:11:43	S-IVB Cutoff - TB5
00:11:53	Earth Parking Orbit Insertion
00:13:00	S-IVB Restart Preparations - TB6
02:33:26	S-IVB Ignition (Translunar Injection)
02:38:48	S-IVB Cutoff - TB7
03:00:00	CSM/S-IVB Separation, SLA Panel Jettison
03:10:00	CSM Turnaround and Dock
04:09:00	CSM/LM Ejection from S-IVB
04:29:00	CSM/LM SPS Evasive Maneuver
04:39:00	S-IVB Slingshot Maneuver
11:33:00	Midcourse Correction - I (SPS)
12:55:00	Crew Rest (9 hours)
26:33:00	Midcourse Correction - 2
27:15:00	TV Transmission to Goldstone
34:00:00	Crew Rest (9 hours)
53:45:00	Midcourse Correction - 3
54:00:00	TV Transmission to Goldstone (15 minutes)
58:00:00	Crew Rest (10 hours)
70:45:00	Midcourse Correction - 4
72:20:00	TV Transmission to Goldstone (15 minutes)
75:45:00	Lunar Orbit Insertion - I
80:10:00	Lunar Orbit Insertion - 2 (Circularization)
80:45:00	TV Transmission to Goldstone (10 minutes)
84:40:00	Crew Rest (8 hours)
94:25:00	Intravehicular Transfer to LM
98:10:00	LM/CSM Separation
98:13:00	TV Transmission to Goldstone (10 minutes)
99:34:00	Descent Orbit Insertion
100:20:00	Landing Radar Operation and Photography
100:46:00	Phasing Maneuver (DPS)
102:33:00	Descent Stage Jettison (LM RCS)
102:43:00	Insertion (APS)
103:34:00	Concentric Sequence Initiation
104:32:00	Constant Delta Height Maneuver
105:09:00	Terminal Phase Initiation
106:20:00	LM/CSM Docking
108:09:00	LM Jettison

Ground Elapsed Time (GET) HR:MIN:SEC:	Event
108:35:00	TV Transmission to Goldstone (15 minutes)
108:39:00	LM APS Burn to Depletion
109:00:00	Crew Rest
119:30:00	Strip Photography (Revolution 23)
122:17:00	Landmark Tracking
126:20:00	TV Transmission to Goldstone (40 minutes)
128:10:00	Lunar Landing Site Photography
128:40:00	Crew Rest (3.5 hours)
137:20:00	Transearth Injection
137:45:00	TV Transmission to Honeysuckle (15 minutes) (Canberra, Australia)
140:30:00	Crew Rest (5.5 hours)
152:20:00	Midcourse Correction - 5
152:35:00	TV Transmission to Goldstone (10 minutes)
154:05:00	Crew Rest (9 hours)
176:50:00	Midcourse Correction - 6
177:30:00	Crew Rest (8 hours)
186:50:00	TV Transmission to Honeysuckle (15 minutes)
188:50:00	Midcourse Correction - 7
191:35:00	CM/SM Separation
191:50:00	Entry Interface
191:50:26	S-band Blackout
191:51:24	Peak G
191:53:26	S-band Blackout Exit
191:58:33	Drogue Parachute Deployment
191:59:22	Main Parachute Deployment
192:04:00	Touchdown

* Based on MSC Revision I of Spacecraft Operational Trajectory for Apollo 10, 28 April 1969, and Saturn V AS-505 Apollo 10 Mission LV Operational Trajectory, 17 April 1969.

Shortly after the spacecraft evasive maneuver, any available residual stage propellants and the Auxiliary Propulsion System (APS) of the S-IVB will be used to perform a retrograde maneuver to reduce the possibility of S-IVB contact with the spacecraft, earth, or moon. S-IVB stage safing will subsequently be performed.

Second Period of Activity

Passive thermal control will be initiated after the first MCC and will be maintained throughout the translunar coast phase unless interrupted by a subsequent MCC. The constraints influencing the translunar coast attitude timeline are thermal control, communications, crew rest cycle, and preferred times of MCC. The translunar coast phase will span approximately 73 hours.

The SPS will be used to insert the docked spacecraft into lunar orbit as shown in Figure 8. The lunar insertion orbit altitude will be approximately 60 x 170 nm. Following insertion, approximately two revolutions in the 60 x 170 nm orbit, and navigation update, the orbit will be circularized at 60 nm. The SPS burn will be initiated near pericynthion of the second revolution. After circularization of the lunar orbit, some LM housekeeping will be accomplished. Subsequently, a simultaneous rest and eat period of approximately 8 hours will be provided for the three astronauts prior to checkout of the LM.

Third Period of Activity

The Commander and Lunar Module Pilot will enter the LM, perform a thorough check of all systems, and

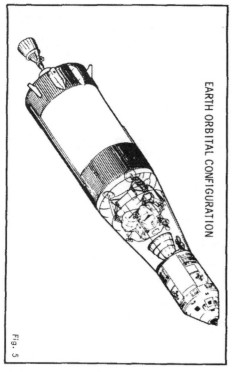

EARTH ORBITAL CONFIGURATION

Fig. 5

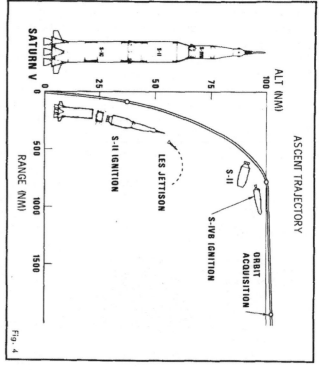

SATURN V

ASCENT TRAJECTORY

ALT (NM)

LES JETTISON

S-II IGNITION

S-II

S-IVB IGNITION

ORBIT ACQUISITION

RANGE (NM)

Fig. 4

TRANSLUNAR CONFIGURATION

Fig. 7

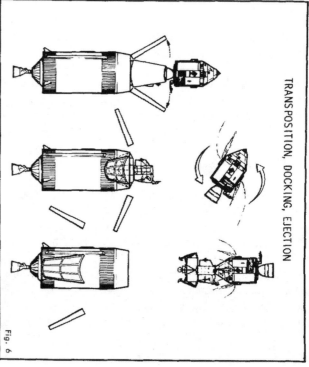

TRANSPOSITION, DOCKING, EJECTION

Fig. 6

undock from the Command/Service Module. The Service Module Reaction Control System (SM RCS) will be used to separate about 30 feet from the LM. Station keeping will be initiated at this point while the Command Module Pilot in the CSM visually inspects the LM. The SM RCS will then be used to perform a separation maneuver directed radially downward toward the moon's center. This maneuver provides a LM/CSM separation at Descent Orbit Insertion (DOI) of about 1.8 nm. The DOI will be performed by a LM Descent Propulsion System (DPS) burn (horizontal, retrograde) (see Figure 9), such that the resulting pericynthion (50,000-foot altitude) occurs about 150 prior to the landing site — the position at which the powered descent is initiated in the Apollo 11 Mission. However, one of the major goals of the Apollo 10 Mission is to accomplish a fidelity demonstration of all phases of a lunar landing mission, except those directly involving LM-powered descent and ascent and lunar surface activities. Relative to the LM-active phase, this goal will be accomplished by incorporating, between the DOI and in-orbit ascent, approximately one phasing revolution during which the required adjustment in CSM lead angle is made (see Figure 9). Near lunar surface activities performed prior to phasing are shown in Figure 10.

LUNAR ORBIT INSERTION

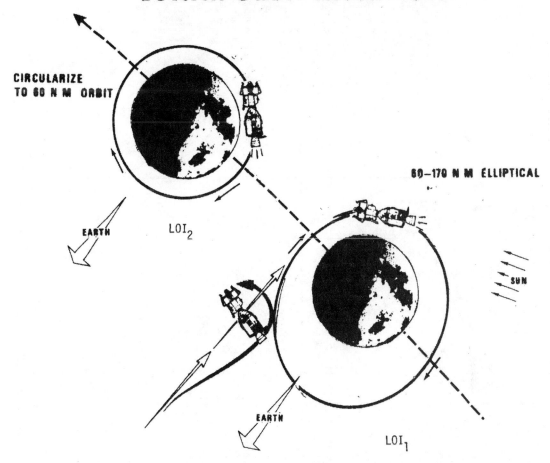

This second LM maneuver will be a DPS burn (posigrade) designed to establish at the resulting LM pericynthion a CSM lead angle equivalent to that which occurs at nominal powered ascent cutoff in a lunar landing mission. This second maneuver is referred to as "phasing." The apocynthion altitude of the phasing orbit will be about 194 nm which will afford the required catch-up time between phasing and the resulting pericynthion of approximately 60,000 feet.

Just prior to this resulting pericynthion, the LM Descent Stage will be jettisoned. Then at pericynthion, a LM Ascent Propulsion System (APS) maneuver (retrograde) will be performed to establish the equivalent of the standard LM insertion orbit (10 by 45 nm) of a lunar landing mission (see Figure 9). At completion of this maneuver, referred to as "insertion" the conditions will be essentially equivalent to those at powered ascent

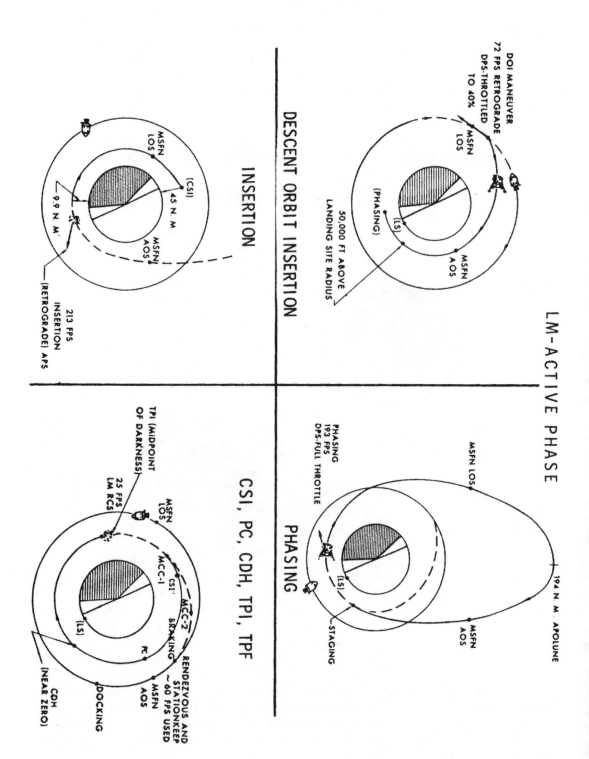

Fig 9.

NEAR LUNAR SURFACE ACTIVITY

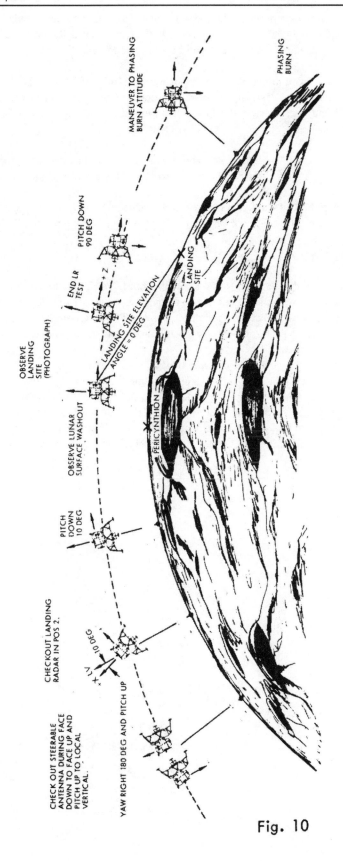

Fig. 10

cutoff for a lunar landing mission. The following maneuvers (see Figure 9), will occur during the rendezvous: concentric sequence initiation (CSI), plane change (PC), constant delta height (CDH), terminal phase initiation (TPI), terminal phase finalization (TPF), and docking. The PC and CDH will nominally be zero.

The LM will coast from insertion to an elliptical orbit (10 by 45 nm) for about an hour. CSI will be initiated at the apocynthion. The terminal maneuver will occur at the midpoint of the period of darkness. After the TPI maneuver and coast period, the LM-to-CSM range will be about 1 nm. Braking during the TPF will be performed manually.

During the LM-active phase discussed above, the CSM will maintain communications with the LM when line-of-sight exists and monitor CSM systems to assure a state-of-readiness if rescue of the LM is required.

Once docked to the CSM, the two LM crewmen will transfer with the exposed film packets and Hasselblad camera to the CSM. The CSM will be separated from the LM using the SM RCS.

Fourth Period of Activity

Following the manned LM activities described above, a LM APS burn to depletion will be commanded by the MSFN in conjunction with the ascent engine arming assembly. Targeting for this burn will be selected to avoid spacecraft recontact, earth return and impact, and lunar impact, respectively, in descending order of priority. Targeting will also be selected to optimize communications during and after the depletion burn.

An 8-hour rest period will ensue, followed by four revolutions to conduct strip photography of Apollo Sites 1 and 2 and to track four well-spaced lunar landmarks. (Figure 11 illustrates tracking of a typical lunar landmark.) Another 4-hour rest period will be followed by about two revolutions of landmark tracking and photography of Site 3. The SPS will be used to inject the CSM into the transearth trajectory after a total time in lunar orbit of about 61.5 hours. The transearth injection burn is currently planned for a nominal transearth return time of 53 hours. The spacecraft transearth configuration is shown in Figure 12.

CSM LUNAR LANDMARK TRACKING

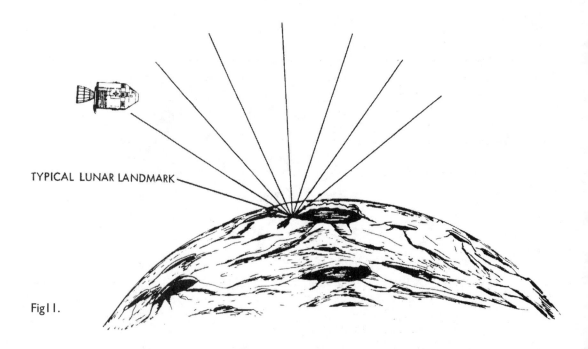

TYPICAL LUNAR LANDMARK

Fig11.

TRANSEARTH CONFIGURATION

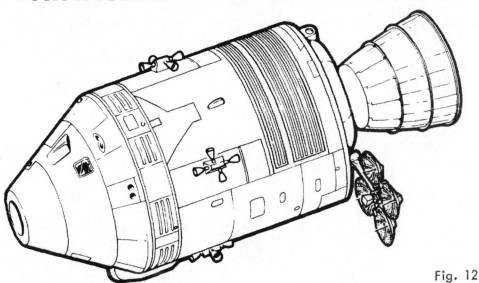

Fig. 12

Fifth Period of Activity

During transearth coast, intermediate MCC's will be made, if required, as shown in Figure 13. These corrections will utilize the MSFN for navigation. In the transearth phase there will be continuous communications coverage from the time the spacecraft appears from behind the moon until about 1 minute prior to entry. The constraints influencing the spacecraft attitude timeline are thermal control, communications, crew rest cycle, and preferred times of MCC's. The attitude profile for the transearth phase is complicated by more severe fuel slosh problems than for the other phases of the mission.

Sixth Period of Activity

Prior to atmospheric entry, the final MCC will be made and the CM will be separated from the SM using the SM RCS. The spacecraft will reach the entry interface (EI) at 400,000 feet, as shown in Figure 14. The S-band communication blackout will begin 26 seconds later followed by C-band communication blackout 30 seconds from EI. The rate of heating will reach a maximum 1 minute 10 seconds after entry. The spacecraft will exit from C-band blackout 3 minutes 4 seconds after entry and from S-band blackout 3 minutes 30 seconds after entry. Drogue parachute deployment will occur 8 minutes 32 seconds after entry at an altitude of 23,300 feet, followed by the main parachute deployment at 9 minutes 20 seconds. Splashdown will occur at entry plus approximately 14 minutes 19 seconds, 1285 nm from EI.

Earth touchdown will be in the Pacific at 165°W longitude, 15°S latitude and will occur 8 days 4 minutes after launch.

CONTINGENCY OPERATIONS

GENERAL

If an anomaly occurs after lift-off that would prevent the space vehicle from following its nominal flight plan, an abort or an alternate mission will be initiated. Aborts will provide for an acceptable flight crew and CM recovery while alternate missions will attempt to maximize the accomplishment of mission objectives as well as providing for an acceptable flight crew and CM recovery.

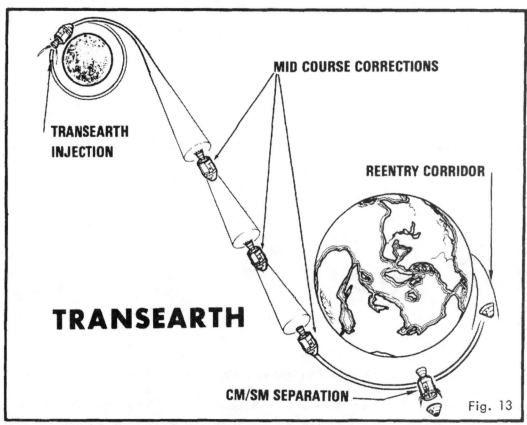

MID COURSE CORRECTIONS

TRANSEARTH
INJECTION

REENTRY CORRIDOR

TRANSEARTH

CM/SM SEPARATION

Fig. 13

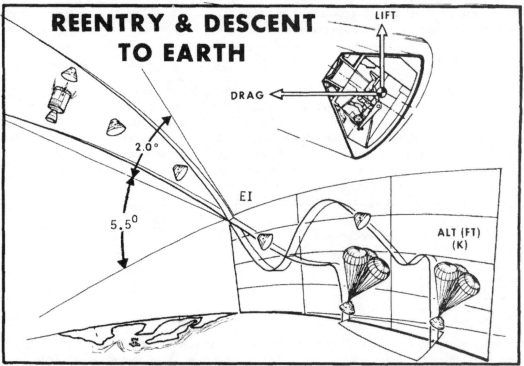

REENTRY & DESCENT
TO EARTH

LIFT

DRAG

2.0°

5.5°

EI

ALT (FT)
(K)

Fig. 14

ABORTS

The following sections describe the abort procedures that may be used to safely return the CM to earth following emergencies that would prevent the space vehicle from following its normal flight plan. The abort descriptions are presented in the order of mission phase in which they could occur.

Launch

There are four different launch abort modes. The following descriptions of the four modes are based on aborts initiated from the nominal launch trajectory. Aborts from a dispersed trajectory will consist of the some procedures, but the times at which the various modes become possible and the resultant landing points may vary.

Mode I - The Mode I abort procedure is designed for safe recovery of the CM following aborts initiated between Launch Escape System activation, approximately 40 minutes prior to lift-off, and Launch Escape Tower jettison, approximately 3 minutes Ground Elapsed Time (GET). The procedure would consist of the Launch Escape Tower pulling the CM off the space vehicle and propelling it a safe distance downrange. The resulting landing point would lie between the launch site and approximately 520 nm downrange.

Mode II - The Mode II abort could be performed from the time the Launch Escape Tower is jettisoned until the full-lift CM landing point is 3200 nm downrange, approximately 10 minutes GET. The procedure would consist of separating the CSM from the launch vehicle, separating the CM from the SM, and then letting the CM free-fall to entry. The entry would be a full-lift, or maximum range trajectory, with a landing on the ground track 440 to 3200 nm downrange.

Mode III - The Mode III abort procedure could be performed from the time the full lift landing point range reaches 3200 nm until orbital insertion. The procedure would consist of separating the CSM from the launch vehicle and then, if necessary, performing a retrograde burn with the SPS so that the half-lift landing point is no farther than 3350 nm downrange. A half-lift entry would be flown which causes the landing point to be approximately 70 nm south of the ground track between 3000 and 3350 nm downrange.

Mode IV and Apogee Kick - The Mode IV abort procedure is an abort to earth parking orbit and could be performed any time after the SPS has the capability to insert the CSM into orbit. This capability begins at approximately 8 minutes 40 seconds GET. The procedure would consist of separating the CSM from the launch vehicle and, shortly afterwards, performing a posigrade SPS burn to insert the CSM into earth orbit. This means that any time during the S-IVB burn portion of the launch phase the CSM has the capability to insert itself into orbit if the S-IVB should fail. The CSM could then remain in earth orbit to carry out an alternate mission, or, if necessary, return to the West Atlantic or Mid-Pacific Ocean after one revolution. The Mode IV abort to orbit capability occurring prior to S-IVB ignition time is a significant change from the previous Saturn V launch abort procedures. This mode of abort is preferred over the Mode II or Mode III aborts and would be used unless an immediate return to earth is necessary during the launch phase. Apogee kick is a variation of the Mode IV abort wherein the SPS burn to orbit would be performed at, or near, first apogee. The main difference between the two is the time at which the posigrade SPS burn is performed.

S-IVB Early Staging to Orbit - Under normal conditions the S-IVB is inserted into orbit with enough fuel to perform the TLI maneuver. This capability can be used, if necessary, during the launch phase to insure that the spacecraft is inserted into a safe parking orbit. Assuming a nominal launch trajectory until 6 minutes GET, the S-IVB then has the capability to be early staged off the S-II and achieve orbit. This means that any time after 6 minutes GET an S-II launch vehicle failure would probably not commit the CSM to a launch abort into the Atlantic. It is preferable to go to earth orbit, if possible, rather than perform a launch abort.

Earth Parking Orbit

Once the S-IVB/IU/LM/SLA/CSM is safely inserted into earth parking orbit, as in the nominal mission, a return-to-earth abort would be performed by separating the CSM from the remainder of the vehicle and then utilizing the SPS for a retrograde burn to place the CM on an atmosphere intersecting trajectory. After entry the crew would fly a guided entry to a pre-selected target point, if available. This procedure would be similar to the deorbit and entry procedure performed on Apollo 7 and 9.

Translunar Injection

Ten-Minute Abort - There is only a remote possibility that an immediate return to earth will become necessary during the relatively short period of the TLI maneuver. However, if it should become necessary the S-IVB burn would be cut off early and the crew would initiate an onboard calculated retrograde SPS abort burn. The SPS burn would be performed approximately 10 minutes after TLI cutoff and would ensure a safe CM entry. The elapsed time from abort initiation to landing would vary from approximately 25 minutes to 4 hours, depending on the length of the TLI maneuver performed prior to S-IVB cutoff. For aborts initiated during the latter portion of TLI, a second SPS burn called a midcourse correction would be necessary to correct for dispersed entry conditions. Since this abort would be used only in extreme emergencies with respect to crew survival, the landing point would not be considered in executing the abort. No meaningful landing point predictions can be made because of the multiple variables involved including launch azimuth, location of TLI, the duration of the TLI burn prior to cutoff, and execution errors of the abort maneuvers.

Ninety-Minute Abort - A more probable situation than the previous case is that the TLI maneuver would be completed and then the crew would begin checking any malfunctions that may have been evident during the burn. If, after the check, it becomes apparent that it was necessary to return to earth, an abort would be initiated at approximately TLI cutoff plus 90 minutes. Unlike the previous procedure, this abort would be targeted to a pre-selected landing location called a recovery line. There are five recovery lines spaced around the earth: one is located in the Atlantic Ocean, one in the Indian Ocean, and three in the Pacific Ocean. The location of these lines is shown in Figure 20. If possible, the abort would be targeted to the Atlantic Ocean recovery line but for some time-critical situations, the abort could be targeted to the East Pacific line. The abort maneuver would be a retrograde SPS burn followed by a midcourse correction, if necessary, performed near apogee to provide the proper CM entry conditions.

Translunar Coast

For approximately 3 days the CSM will be in the translunar coast (TLC) phase of the mission. The abort procedure during this time would be similar to the 90-minute abort. Abort information specifying a combination of SPS burn time and CSM attitude would be sent to the crew to be performed at a certain time. All aborts initiated during translunar coast will return the CM to approximately the some inertial point in space where the TLI maneuver was performed. Therefore, since the point where the CM will contact the atmosphere is inertially fixed and the earth is rotating, the latitude of landing is constant but the longitude is changing 15° every hour, a movement equal to the earth's rate of rotation. This means that each of the five recovery lines pass through the predicted landing point once every 24 hours. The landing longitude is controlled by selecting an abort trajectory that causes the CM to enter the atmosphere and land at the time the target longitude is passing near the inertial point where TLI was performed.

Fixed times of abort that will return the CM to the Mid-Pacific recovery line will be selected during TLC. But, since the Mid-Pacific recovery line passes under the inertial position of TLI only once every 24 hours, a landing on this line can occur only once every 24 hours. A time-critical situation may dictate targeting the abort to one of the other four lines to decrease the elapsed time from abort to landing. Deep space aborts after TLI plus 90 minutes would be targeted to, in order of priority, (1) the Mid-Pacific line, (2) the Atlantic Ocean line, (3) the East Pacific line, the West Pacific line, or the Indian Ocean line. Regardless of the recovery line selected, the landing latitude should remain nearly the some. The minimum elapsed time between abort initiation and CM landing increases with translunar coast flight time. About the time the CSM enters the moon's sphere of gravitational influence, it becomes faster to perform a circumlunar abort rather than returning directly to earth.

Lunar Orbit Insertion

Aborts following an early shutdown of the SPS during the lunar orbit insertion (LOI) maneuver are divided into three categories, Mode I, II, and III. All aborts performed during LOI will return the CM to the latitude of the moon's antipode at the time of the abort maneuver. The longitude will depend on the return flight time but will normally be the Mid-Pacific recovery line.

Mode I - The Mode I procedure would be used for aborts following SPS cutoffs from ignition to approximately 2 minutes into the LOI burn. This procedure would consist of performing a posigrade DPS burn approximately 2 hours after cutoff to put the spacecraft back on a return-to-earth trajectory.

Mode II - The Mode II procedure would be used for aborts following shutdown between LOI ignition plus 2 minutes and LOI ignition plus 3 minutes. The abort maneuver is performed in two stages. The first DPS burn would be done to reduce the lunar orbital period and to insure that the spacecraft does not impact the lunar surface. After one orbit a second DPS burn would be performed near pericynthion to place the spacecraft on a return-to-earth trajectory targeted to the mid-Pacific recovery line.

Mode III - The Mode III procedure would be used for aborts following shutdowns from approximately 3 minutes into the burn until nominal cutoff. After 3 minutes of LOI burn, the spacecraft will have been inserted into an acceptable lunar orbit. Therefore, the abort procedure would be to let the spacecraft go through one or two lunar revolutions prior to doing a posigrade DPS burn at pericynthion. This would place the spacecraft on a return-to-earth trajectory targeted to the MidPacific recovery line.

Lunar Orbit

Aborts from the lunar orbit would be accomplished by performing the transearth injection (TEI) burn early. The abort would be targeted to the Mid-Pacific recovery line.

Transearth Injection

The abort procedures for early cutoff of TEI are the inverse of the LOI abort procedures except that the maneuver would be performed with the SPS instead of the DPS and no Mode II abort would be necessary. This is due to the fact that jettisoning the LM reduces the weight of the spacecraft enough to allow a direct changeover from Mode III to Mode I. That is, for early cutoffs between TEI ignition and approximately 2 minutes, a Mode III abort would be performed. After this time a Mode I abort would be used. All TEI aborts should result in landings on the Mid-Pacific recovery line at the latitude of the moon's antipode at TEI.

Transearth Coast

From TEI until entry minus 24 hours, the only abort procedure that could be performed is to use the SPS or the SM RCS for a posigrade or retrograde burn that would respectively decrease or increase the transearth flight time and change the longitude of landing. After entry minus 24 hours, no further burns to change the landing point will be performed. This is to ensure that the CM maintains the desired entry velocity and flight path angle combination that will allow a safe entry.

Entry

If, during entry, the Guidance, Navigation, and Control System (GN&CS) fails, a guided entry to the end-of-mission target point cannot be flown. In this case, the crew would use their Entry Monitoring System (EMS) to fly a 1285-nautical mile range. The landing point would be abeam the guided entry target point on the north side of the ground track. If both the GN&CS and EMS fail, a "constant g" (constant deceleration) entry would be flown. The landing point would be approximately 185 nm uprange of the guided target point and 75 nm north of the ground track.

ALTERNATE MISSION SUMMARY

Consideration of previous mission accomplishments will aid in determining which available alternate would be most appropriate to implement in the event of certain Apollo 10 Mission contingencies.

Launch Vehicle Alternates

Alternate I

Condition/Malfunction:
>   A. One S-IC stage engine out; or
>   B. One S-II stage engine out; or
>   C. Early staging of S-IVB from S-II

Perform: Nominal lunar landing mission TLI if capability exists.

Alternate 2

Condition/Malfunction:
>   Same as Alternate 1.

Perform: Spacecraft alternate mission consistent with real-time evaluation of capability.

Alternate 3

Condition/Malfunction:
>   D. Early shutdown of S-IVB engine during first burn.

Perform: SPS earth parking orbit insertion.

Alternate 4

Condition/Malfunction:
>   E. S-IVB engine inhibited for first opportunity TLI burn.

Perform: Second TLI opportunity restart, and nominal lunar mission if restart successful.

Alternate 5

Condition/Malfunction:
>   F. S-IVB fails to restart for TLI; or
>   G. S-IVB fails to reach TLI velocity; or
>   H. S-IVB inhibited for second TLI opportunity.

Perform: TD&E, propellant dumping, S-IVB APS ullage motor firing, and stage safing.

Spacecraft Earth Orbit Alternates

Alternate I - CSM-Only Low Earth Orbit

Condition/Malfunction:
>   LM not ejected, or S-IVB failed prior to 25,000-nm apogee; or SPS used to achieve earth orbit.

Perform: SPS LOI simulation (100 x 400-NM orbit), MCC's to approximate lunar timeline and for an approximate 10-day mission with landing in 150°W Pacific recovery area.

Alternate 2 - CSM-Only Semisynchronous

Condition/Malfunction:
     S-IVB fails during TLI with apogee greater than or equal to 25,000 nm, LM cannot be ejected.

Perform: SPS phasing maneuver for LOI tracking, LOI simulation, SPS phasing maneuver to place perigee over Pacific recovery zone at later time, SPS semisynchronous orbit, and further MCC's to approximate lunar timeline.

Alternate 3 - CSM/LM Earth Orbit Combined Operations with SPS Deboost

Condition/Malfunction:
     TLI does not occur or TLI apogee less than 4000 nm, TD&E successful.

Perform: SPS maneuver to raise or lower apogee for lifetime requirements if necessary, simulated LOI to raise or lower apogee to 400 nm, simulated DOI (in docked configuration), simulated powered descent insertion (PDI), SPS maneuver to circularize at 150 nm, LM-active rendezvous, APS burn to depletion (unmanned, Abort Guidance System (AGS)-controlled), and further SPS MCC's to complete lunar mission timeline.

Alternate 4 - CSM/LM Earth Orbit Combined Operations with DPS/SPS Deboost

Condition/Malfunction:
     S-IVB fails during TLI, SPS and DPS in combination can return CSM/LM to low earth orbit without sacrificing LM rescue (apogee less than 10,000 nm but greater than 4000 nm).

Perform: SPS phasing maneuver, simulated DOI, PDI to lower apogee to about 4000 nm, SPS phasing (simulated MCC) maneuver to ensure tracking for LOI, SPS maneuver to circularize at 150 nm, LM-active rendezvous, APS burn to depletion (unmanned, AGS-controlled), SPS maneuvers to complete lunar mission timeline, and achieve nominal 90 x 240 nm end of mission orbit for an approximate 10-day mission with landing on 150°W Pacific recovery area.

Alternate 5 - CSM/LM Semisynchronous

Condition/Malfunction:
     SPS and DPS in combination cannot place CSM/LM in low earth orbit without sacrificing LM rescue, SPS propellant not sufficient for CSM/LM circumlunar mission.

Perform: SPS phasing maneuver (to place a later perigee over an MSFN site), SPS LOI (approximately semisynchronous), SPS phasing maneuver if necessary to adjust semisynchronous orbit, docked DPS DOI, docked DPS PDI simulation, SPS phasing to put perigee over or opposite recovery zone, SPS to semisynchronous orbit, and further MCC's to approximate lunar mission timeline.

Spacecraft Lunar Alternates

Alternate la - DPS LOI

Condition/Malfunction:
     Non-nominal TLI such that: CSM/LM LOI and TEI No-Go with SPS, CSM/LM LOI Go with DPS LOI-1.

Perform: TD&E, SPS free-return CSM/LM, DPS LOI-1, and SPS LOI-2. After LOI-2, the Descent Stage is jettisoned and the Ascent Stage rendezvous is performed.

Alternate 1b - CSM Solo Lunar Orbit

Condition/Malfunction:
        Non-nominal TLI such that: CSM/LM LOI No-Go, CSM only LOI Go.

Perform: TD&E, SPS free-return CSM/LM LM testing during TLC, DPS staging, unmanned APS depletion burn during TLC, and CSM lunar mission (Alternate 2).

Alternate 1c - CSM/LM Flyby

Condition/Malfunction:
        Non-nominal TLI, such that: CSM/LM Flyby Go, CSM/LM LOI No-Go, CSM only LOI No-Go.

Perform: Transposition, docking, and ejection; LM testing near pericynthion, docked DPS maneuver to raise pericynthion, DPS staging, unmanned APS depletion burn, and SPS for fast return.

Alternate 2 - CSM-Only Lunar Orbit

Condition/Malfunction:
        Failure to TD&E

Perform: CSM-only lunar orbit mission, landmark tracking, lunar surface photography, and low pericynthion MSFN tracking. Follow lunar landing mission work-rest cycle.

Alternate 3a - DPS TEI and/or APS Depletion

Condition/Malfunction:
        LM No-Go for undocking and rendezvous, but DPS Go for a burn.

Perform: Landmark tracking, DPS TEI, unmanned APS depletion burn, and SPS maneuver for fast return.

Alternate 3b - APS Depletion in Lunar Orbit

Condition/Malfunction:
        LM No-Go for undocking, DPS No-Go for a burn.

Perform: Simulated nominal timeline closely until nominal time for phasing, two additional revolutions of LM testing, unmanned APS depletion burn, landmark tracking. Revert to nominal timeline.

Alternate 4 - TEI with Docked Ascent Stage

Condition/Malfunction:
        CSM communication failure in lunar orbit.

Perform: TEI and keep LM as communication system. If DPS available, perform DPS TEI as in Alternate 3. If Descent Stage jettisoned, perform SPS TEI with Ascent Stage attached.

Spacecraft Rendezvous Alternates

Alternate 1a - Descent Stage - Unstaged

Condition/Malfunction.
        Descent Stage cannot be jettisoned.

Perform: CSM minifootball (SM RCS), DOI (DPS), phasing (DPS), insertion (DPS), CSI (DPS), CDH (DPS), TPI (DPS), and TPF (SM RCS).

Fig. 15

Launch Escape System

Command Module

Service Module

Lunar Module

Instrument Unit

Fuel Tank

LOX Tank

J-2 Engine (1)

Fuel Tank

LOX Tank

J-2 Engines, (5)

LOX Tank

Fuel Tank

F-1 Engines, (5)

22' Diam

S-IV B Stage 59'

~ 363'

S-II Stage 81'

33' Diam

S-IC Stage 138'

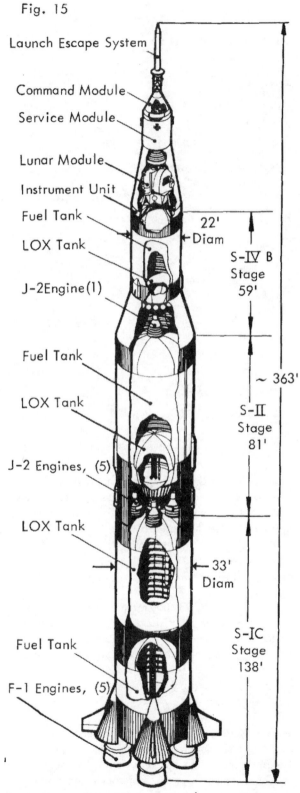

APOLLO/SATURN V SPACE VEHICLE

Alternate 2 - APS Rendezvous

Condition/Malfunction:
    DPS cannot be used.

Perform: CSM minifootball (SM RCS), Descent Stage jettison (LM, RCS), DOI to about 58-NM pericynthion (APS), phasing (APS), CSI (LM RCS, interconnect), CDH (APS), and TPI and TPF (LM, RCS).

Alternate 3 - Modified Football

Condition/Malfunction:
    Unusable DPS and APS - no other rendezvous possible. Perform: Descent Stage jettison (LM RCS), TPI (LM RCS), and TPF (LM RCS).

SPACE VEHICLE DESCRIPTION

The Apollo 10 Mission will be performed by an Apollo/Saturn V Space Vehicle (Figure 15) designated AS-505, which consists of a three-stage Saturn V Launch Vehicle, and a complete Apollo Block 11 Spacecraft. A more comprehensive description of the space vehicle and its subsystems is included in the Mission Operation Report Supplement. The following is a brief description of the various stages of AS-505.

The Saturn V Launch Vehicle (SA-505) consists of three propulsion stages (S-IC, S-II, S-IVB) and an Instrument Unit (IU). The Apollo Spacecraft payload for Apollo 10 consists of a Launch Escape System (LES), Block If Command/Service Module (CSM 106), a Spacecraft LM Adapter (SLA 13), and a Lunar Module (LM-4). A list of current weights for the space vehicle is contained in Table 2.

LAUNCH VEHICLE DESCRIPTION

First Stage (S-IC)

The S-IC is powered by five F-1 rocket engines each developing approximately 1,522,000 pounds of thrust at sea level and building up to 1.7 million pounds before cutoff. One engine, mounted on the vehicle longitudinal centerline, is fixed; the remaining four engines, mounted in a square pattern about the centerline, are gimbal led for thrust vector control by signals from the control system housed in the IU. The F-1 engines utilize LOX (liquid oxygen) and RP-1 (kerosene) as propellants.

TABLE 2

APOLLO 10 WEIGHT SUMMARY

(Weight in Pounds)

STAGE/MODULE	INERT WEIGHT	TOTAL EXPENDABLES	TOTAL WEIGHT	FINAL SEPARATION WEIGHT
S-IC Stage	294,300	4,738,320	5,032,620	368,070
S-IC/S-II Interstage	11,450		11,450	
S-II Stage	84,490	988,480	1,072,970	98,080
S-II/S-IVB Interstage	8,080		8,080	
S-IVB Stage	25,710	236,040	261,750	28,930
Instrument Unit	40,250		4,250	
Launch Vehicle at Ignition (SA-505)	6,391,120			
SCAM Adapter (SLA-13)	4,060		4,060	
Lunar Module (LM-4)	10,170	20,610	30,780	
Service Module (SM-106)	10,160	40,640	51,250	13,160
Command Module (CM-106)	12,250		12,250	11,141 (Splashdown)
Launch Escape System	8,890		8,890	
Spacecraft at Ignition	107,230			
Space Vehicle at Ignition (AS-505)			6,498,350	
S-IC Thrust Buildup			-86,100	
Space Vehicle at Lift-off			61412,250	
Space Vehicle at Earth Orbit Insertion			295,150	

Second Stage (S-II)

The S-II is powered by five high-performance J-2 rocket engines each developing approximately 230,000 pounds of thrust in a vacuum. One engine, mounted on the vehicle longitudinal centerline, is fixed; the remaining four engines, mounted in a square pattern about the centerline, are gimbaled for thrust vector control by signals from the control system housed in the IU. The J-2 engines utilize LOX and LH2 (liquid hydrogen) as propellants.

Third Stage (S-IVB)

The S-IVB is powered by a single J-2 engine developing approximately 230,000 pounds of thrust in a vacuum. As installed in the S-IVB, the J-2 engine features a multiple start capability. The engine is gimbaled for thrust vector control in pitch and yaw. Roll control is provided by the Auxiliary Propulsion System (APS) modules containing motors to provide roll control during mainstage operations and pitch, yaw, and roll control during non-propulsive orbital flight.

Instrument Unit

The Instrument Unit (IU) contains the following: Electrical System, supplies electrical power for all IU system components; Environmental Control System, provides thermal conditioning for the electrical components and guidance systems contained in the assembly; Guidance and Control System, solves guidance equations and controls the attitude of the vehicle; Measuring and Telemetry System, monitors and transmits flight parameters and vehicle operation information to ground stations; Radio Frequency System, provides for tracking and command signals; components of the Emergency Detection System (EDS).

SPACECRAFT DESCRIPTION

Command Module

The Command Module (CM) (Figure 16) serves as the command, control, and communications center for most of the mission. Supplemented by the Service Module, it provides all life support elements for three crewmen in the mission environments and for their safe return to earth's surface. It is capable of attitude control about three axes and some lateral lift translation at high velocities in earth atmosphere. It also permits Lunar Module attachment, Command Module/Lunar Module ingress and egress, and serves as a buoyant vessel in open ocean.

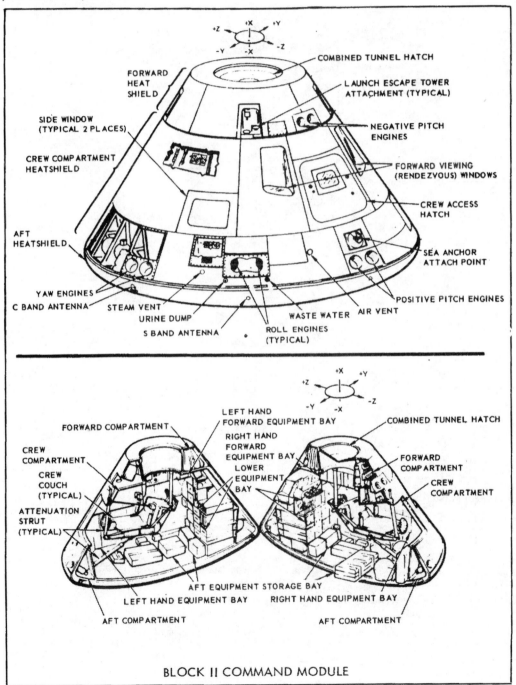

BLOCK II COMMAND MODULE

Fig. 16

Service Module

The Service Module (SM) (Figure 7) provides the main spacecraft propulsion and maneuvering capability during the mission. The Service Propulsion System (SPS) provides up to 20,500 pounds of thrust in a vacuum. The Service Module Reaction Control System (SM RCS) provides for maneuvering about and along three axes. The SM provides most of the spacecraft consumables (oxygen, water, propellant, hydrogen). It supplements environmental, electrical power, and propulsion requirements of the CM. The SM remains attached to the CM until it is jettisoned just before CM entry.

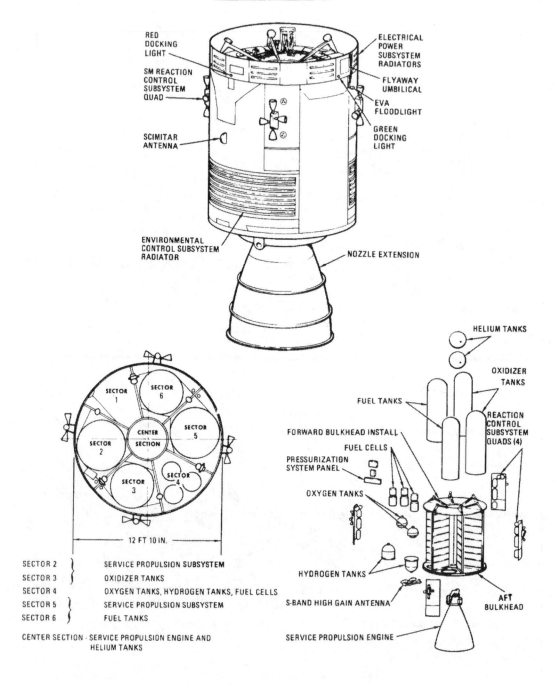

SERVICE MODULE

SECTOR 2   } SERVICE PROPULSION SUBSYSTEM
SECTOR 3   } OXIDIZER TANKS
SECTOR 4   OXYGEN TANKS, HYDROGEN TANKS, FUEL CELLS
SECTOR 5   } SERVICE PROPULSION SUBSYSTEM
SECTOR 6   } FUEL TANKS

CENTER SECTION - SERVICE PROPULSION ENGINE AND
                 HELIUM TANKS

Fig. 17

Common Command/Service Module Systems

There are a number of systems which are common to the CSM.

Guidance and Navigation System

The Guidance and Navigation (G&N) System measures spacecraft attitude and acceleration, determines trajectory, controls spacecraft attitude, controls the thrust vector of the SPS engine, and provides abort information and display data.

Stabilization and Control System

The Stabilization and Control System (SCS) provides control and monitoring of the spacecraft attitude, backup control of the thrust vector of the SPS engine and a backup inertial reference.

Reaction Control Systems

The reaction control systems (RCS) provide thrust for attitude and small translational maneuvers of the spacecraft in response to automatic control signals from the SCS in conjunction with the G&N system. The CM and SM each has its own independent and redundant system, the CM RCS and the SM RCS respectively. Propellants for the RCS are hypergolic.

Electrical Power System

The Electrical Power System (EPS) supplies all electrical power required by the CSM. The primary power source is located in the SM and consists of three fuel cells which are the prime spacecraft power from lift-off through CM/SM separation. Five batteries — three for peak load intervals, entry and post-landing, and two for pyrotechnic uses — are located in the CM.

Environmental Control System

The Environmental Control System (ECS) provides a controlled cabin environment and dispersion of CM equipment heat loads.

Telecommunications System

The Telecommunications (T/C) System provides for the acquisition, processing, storage, transmission and reception of telemetry, tracking, and ranging data among the spacecraft and ground stations.

Sequential System

Major Sequential Subsystems (SEQ) are the Sequential Events Control System (SECS), Emergency Detection System (EDS), Launch Escape System (LES), and Earth Landing System (ELS). The subsystems interface with the RCS or SPS during an abort.

Spacecraft LM Adapter

The Spacecraft LM Adapter (SLA) is a conical structure which provides a structural load path between the LV and SM and also supports the LM. Aerodynamically, the SLA smoothly encloses the SM engine nozzle and irregularly-shaped LM, and transitions the SV diameter from that of the upper LV stage to that of the SM. The upper section is made up of four panels that swing open at the top and are jettisoned away from the spacecraft by springs attached to the lower fixed panels.

Lunar Module

The Lunar Module (LM) (Figure 18) is a two-stage vehicle designed to transport two crewmen from a docked position with the CSM to the lunar surface, serve as a base for lunar surface crew operations, and provide for their safe return to the docked position. The upper stage is termed the Ascent Stage (AS) and the lower stage, the Descent Stage (DS). In the nominal mission, the two stages are operated as a single unit until the lunar landing is accomplished. The AS is used for ascent from the lunar surface and rendezvous with the CSM.

LUNAR MODULE

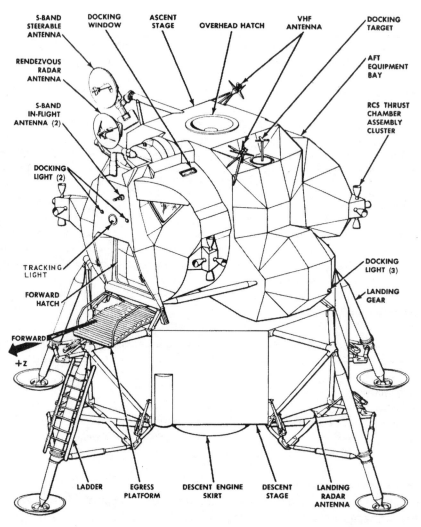

Fig. 18

The LM's main propulsion includes a gimbaled, throttleable Descent Propulsion System (DPS) engine and a fixed, non-throttleable Ascent Propulsion System (APS) engine. A 16-jet LM Reaction Control System (LM RCS) on the AS provides for stabilization and maneuvering. All propulsive systems utilize storable hypergolic propellants. The Guidance, Navigation, and Control System (GN&CS) has the capability to implement automatically all parameters required for safe landing from lunar orbit and accomplish a CSM/LM rendezvous from lunar launch. Landing and rendezvous radar systems aid the GN&CS system. The Instrumentation System (IS) provides for LM systems checkout and displays data for monitoring or manually controlling LM systems. The ECS provides a satisfactory environment for equipment and human life. The EPS relies upon four batteries in the DS and two batteries in the AS when undocked from the CSM. Electrical power is provided by the CSM when the LM is docked. Telecommunications is provided to the MSFN and the CSM.

Cameras

Hasselblad Cameras

One Hasselblad camera will be carried in the CM and in the LM. Lens provisions will include an 80mm lens for each camera and a 250mm lens will be carried in the CM.

The Hasselblad cameras will be used for long distance earth and lunar terrain/surface photography as well as coverage of LM-active rendezvous and TD&E.

Data Acquisition Cameras

One 16mm Maurer (Data Acquisition) camera will be carried in the CM and in the LM. Maurer camera provisions will include 5, 18, and 75mm lenses.

The data acquisition cameras will be used to photograph crew activities, star/horizon/ lunar landmarks, LM inspection, and to cover LM-active rendezvous and TD&E.

Television Cameras

A black and white TV camera (RCA) and a color TV camera (Westinghouse) will be carried in the CM. The cameras will provide realtime views of the lunar surface, the earth from lunar distances, and the crew activities inside the CM.

Launch Escape System

The Launch Escape System (LES) provides the means for separating the CM From the LV during pad or suborbital aborts through approximately one-half minute of the second stage burn. This system consists primarily of the Launch Escape Tower (LET), Launch Escape Motor, Tower Jettison Motor, and Pitch Motor. All motors utilize solid propellants. A Boost Protective Cover (BPC) is attached to the LET and covers the CM from LES rocket exhaust and also from aerodynamic heat generated during LV boost.

CONFIGURATION DIFFERENCES

The space vehicle for Apollo 10 varies in its configuration from that flown on Apollo 9 and those to be flown on subsequent missions because of normal growth, planned changes, and experience gained on previous missions. Following is a list of the major configuration differences between AS-504 and AS-505.

COMMAND/SERVICE MODULE (CM-106)

Added VHF ranging capability as a backup to CSM/LM rendezvous radar (RR).

LUNAR MODULE (LM-4)

Added VHF ranging capability as an RR backup.

Incorporated CM to LM power transfer capability after LM stage separation to extend hold capability between docking and final separation (contingency).

Provided CM/LM, power transfer redundancy as a power transfer backup.

Deleted EVA antenna because no EVA planned for Apollo 10.

Increased digital uplink voice output (up to 20 db) because required for lunar distance communication.

Added landing gear deployment mechanism protective shield to prevent possible malfunction due to DPS plume impingement.

Added AS plume heat blanket and venting to improve thermal control.

Added separate power source for utility/floodlight to prevent simultaneous loss of both lights.

Added APS muffler to prevent APS regulator loss.

Provided RR and VHF bus isolation to prevent simultaneous RR and VHF loss.

Deleted TV camera.

Substituted Luminary I (onboard program).

SLA- 13

(No significant differences.)

INSTRUMENT UNIT (S-IU-505)

Incorporated IU network change (software) to enable SC control of LV during launch phase.

Added damping compound to and removal of channels from ST-124M platform support because of the first time application of this damping approach to IU.

S-IVB STAGE (SA-505)

Substituted new design helium regulator valve because of the first-time flight of new hardware fix - SA-504 malfunction.

S-II STAGE (SA-505)

Planned center engine early cutoff as a possible elimination of longitudinal oscillations.

S-IC STAGE (SA-505)

(No significant differences.)

HUMAN SYSTEM PROVISIONS

The major human system provisions included for the Apollo 10 mission are: Space Suits, Bioinstrumentation System, Medical Provisions, Crew Personal Hygiene, Crew Meals, Sleeping Accommodations, Oxygen Masks, and Survival Equipment. These systems provisions are described in detail in the Mission Operation Report Supplement.

LAUNCH COMPLEX

The AS-505 Space Vehicle (SV) will be launched from Launch Complex (LC) 39, Pad B at the Kennedy Space Center (KSC). The major components of LC 39 include the Vehicle Assembly Building (VAB), the Launch Control Center (LCC), the Mobile Launcher (ML), the Crawler Transporter (C/T), the Mobile Service Structure (MSS), and the Launch Pad.

The LCC is a permanent structure located adjacent to the VAB and serves as the focal point for monitoring and controlling vehicle checkout and launch activities for all Saturn V launches. The ground floor of the structure is devoted to service and support functions. Telemetry equipment occupies the second floor and the third floor is divided into firing rooms, computer rooms, and offices. Firing room 3 will be used for Apollo 10.

The AS-505 SV was received at KSC and assembly and initial overall checkout was performed in the VAB on the mobile launcher. Rollout occurred on 11 March 1969. Transportation to the pad of the assembled SV and ML was provided by the Crawler Transporter (C/T) which also moved the MSS to the pad after the ML and SV had been secured. The MSS provides 360-degree access to the SV at the launch pad by means of five vertically-adjustable, elevator-serviced, enclosed platforms. The MSS will be removed to its park position prior to launch.

The emergency egress route system at LC 39 is made up of three major components: the high speed elevators, slide tube, and slide wire. The primary route for egress from the CM is via the elevators and, if necessary, through the slide tube which exits into an underground blast room.

A more complete description of LC 39 is in the MOR Supplement.

An aerial view of Launch Complex 39 with AS-505 on Pad B is shown in Figure 19.

Fig 19.

AERIAL VIEW OF APOLLO 10 SPACE VEHICLE ON PAD B, LC-39

MISSION SUPPORT

Mission support is provided by the Launch Control Center (LCC), the Mission Control Center (MCC), the Manned Space Flight Network (MSFN), and the recovery forces. The LCC is essentially concerned with prelaunch checkout, countdown, and with launching the SV, while MCC located at Houston, Texas, provides centralized mission control from lift-off through recovery. The MCC functions within the framework of a Communications, Command, and Telemetry System (CCATS); Real Time Computer Complex (RTCC); Voice Communications System; Display/Control System; and, a Mission Operations Control Room (MOCR). These systems allow the flight control personnel to remain in contact with the spacecraft, receive telemetry and operational data which can be processed by the CCATS and RTCC for verification of a safe mission or compute alternatives. The MOCR is staffed with specialists in all aspects of the mission who provide the Mission Director and Flight Director with real time evaluations of mission progress.

The MSFN is a worldwide communications network which is controlled by the MCC during Apollo missions. The network is composed of fixed stations (Figure 25) and is supplemented by mobile stations (Table 3) which are optimally located within a global band extending from approximately 40° south latitude to 40° north latitude. Station capabilities are summarized in Table 4.

The functions of these stations are to provide trucking, telemetry, and command and communications both on an updata link to the spacecraft and on a downdata link to the MCC. Connection between these many MSFN stations and the MCC is provided by NASA Communications Network (NASCOM). More detail on Mission Support is in the MOR Supplement.

TABLE 3

MSFN MOBILE FACILITIES

SHIPS	LOCATION		SUPPORT
USNS Vanguard	25°N	49°W	Insertion
USNS Mercury	32°S	131°E	Injection
USNS Redstone	14°S	145.5°E	Injection
	17°S	169°E	Reentry
USNS Huntsville	21°S	173°W	Reentry (tentative)

APOLLO RANGE INSTRUMENTATION AIRCRAFT (ARIA)

Eight ARIA will be available to support the AS-505 mission. The aircraft will operate in the Pacific or Atlantic sector as appropriate. The mission plan calls for ARIA support of translunar injection (TLI) on revolution 2 or 3 and from reentry (400,000-foot altitude) to recovery of the spacecraft crew after splashdown.

RECOVERY SUPPORT PLAN

GENERAL

The major responsibilities of the recovery forces in supporting this mission include the rapid location and safe retrieval of the flight crew, the CM photographic film and recordings, and return of test data and test hardware. This responsibility commences when the space vehicle leaves the launch pad and ends with safe return of the CM and the flight crew to designated points within the continental United States. Recovery force locations and functions are listed in Table 5.

LAUNCH PHASE

Launch Site Area

The launch site area includes all possible CM landing points which would occur following aborts initiated between Launch Escape System (LES) activation and approximately 90 seconds GET. Planned recovery forces will have a capability to provide a maximum access time of 30 minutes to any point in the area. This support is required from the time the LES is armed (approximately lift-off minus 40 minutes) until 90 seconds after lift-off. However, prior to LES arming, the launch site forces are required to be ready to provide assistance if needed to the Pad Egress Team and, after lift-off plus 90 seconds, they are required to provide assistance to the launch abort area recovery forces.

TABLE 4

NETWORK CONFIGURATION FOR APOLLO 10 MISSION

Facilities	C-band (High Speed)	C-band (Low Speed)	ODOP	Optical	USB	Voice (A/G)	Command	Telemetry	TV	VHF Links	FM Remoting	Mag Tape Recording	Decoms	Displays	CMD Destruct	642B TLM	642B CMD	1218	High Speed Data	Wideband Data	TTY	Voice (SCAMA)	VHF A/G Voice	TV Remoting	SPAN	Riometer
	Tracking				USB					TLM					CMD	Data Processing			Comm						Other	
CIF								X	X	X		X	X						X	X		X				
TEL 4										X		X														
CNV	X		X	X											X											
PAT	X	X																								
MLA	X	X																								
MIL					X	X	X	X	X	X	X	X	X	X		X	X	X	X		X	X	X	X		
GBI										X		X			X											
GBM					X	X	X	X	X	X	X	X	X			X	X	X	X		X	X	X			
GTK	X	X													X											
ANG					X	X	X	X	X	X	X	X	X			X	X	X	X		X	X	X			
ANT	X	X										X														
BDA	X	X			X	X	X	X	X	X	X	X	X	X	X	X	X	X	X		X	X	X			
ACN					X	X	X	X	X	X	X	X	X			X	X	X	X		X	X	X			
ASC	X	X																								
MAD					X	X	X	X	X		X	X	X			X	X	X	X		X	X			X	
MADX					X	X	X	X	X									X			X	X				
CYI					X	X	X	X	X	X	X	X	X			X	X	X	X		X	X	X		X	X
PRE		X																								
TAN		X								X		X									X	X	X			
CRO	X	X			X	X	X	X	X	X	X	X	X			X	X	X	X		X	X	X		X	X
HSK					X	X	X	X	X		X	X	X			X	X	X	X		X	X				
HSKX					X	X	X	X	X									X			X	X				
GWM					X	X	X	X	X	X	X	X	X			X	X	X	X		X	X	X			
HAW	X	X			X	X	X	X	X	X	X	X	X	X		X	X	X	X		X	X	X			
CAL	X	X																			X	X	X			
GDS					X	X	X	X	X		X	X	X			X	X	X	X		X	X		X		
GDSX					X	X	X	X	X									X			X	X				
GYM					X	X	X	X	X	X	X	X	X			X	X	X	X		X	X	X			
TEX					X	X	X	X	X	X	X	X	X			X	X	X	X		X	X	X			
HTV		X			X	X		X		X		X	X								X	X	X			
RED	X	X			X	X	X	X	X	X	X	X	X			X	X			X	X	X	X			
VAN	X	X			X	X	X	X	X	X	X	X	X			X	X				X	X	X			
MER	X	X			X	X	X	X	X	X	X	X	X			X	X				X	X	X			
ARIA (6)					X	X		X		X		X										X	X			
LIMA (Peru)																										X

Launch Abort Area

The launch abort area is where the CM will land following an abort initiated during the launch phase of flight. The launch abort area is divided into two sectors, A and B. These sectors are used to differentiate between the recovery force support required in the eastern and western portions of the area. Sector A is all the area in the launch abort area that is within 1000 nautical miles of the launch site. This sector includes the area where the CM would land following aborts initiated prior to the time the launch vehicle achieves early S-IVB staging to orbit capability. Sector B is all the remaining area in the launch abort area.

Recovery forces have the capability to provide a maximum access time of 4 hours to any point on the spacecraft ground track within the area, a maximum CM and crew retrieval time of 24 hours to any point in Sector A, and a maximum CM and crew retrieval time of 48 hours to any point on the spacecraft ground track for a 72° launch azimuth in Sector B. Landings in the remainder of Sector B are considered to be contingency landings with no maximum retrieval time defined.

Table 5

RECOVERY FORCE LOCATIONS, APOLLO 10

Nominal Mission Phase Ships

Designation	Location	Mission Coverage
Primary Recovery Ship (PRS) USS Princeton	24°00'S latitude 165°00'W longitude 15°08'S latitude 165°00'W longitude	TLC to LOI minus 10 hours Entry minus 6 hours to landing
Secondary Recovery Ship (SRS) 1 USS Rich	28°00'N latitude 70°00'W longitude	Launch window opening to earth parking orbit insertion.
Secondary Recovery Ship (SRS) 2 USS Vanguard	25°00'N latitude 49°00'W longitude	Launch window opening to earth parking orbit insertion.
Secondary Recovery Ship (SRS) 3 USS Chilton	30°00'N latitude 38°30'W longitude	Launch window opening to earth parking orbit insertion and stand by for possible earth orbital alternate mission.
Secondary Recovery Ship (SRS) 4 USS Ozark	25°00'S latitude 25°00'W longitude 15°08'S latitude 25°00'W longitude	TLC to LOI minus 10 hours Entry minus 6 hours to landing
Secondary Recovery Ship (SRS) 5 USS Carpenter	21°45'N latitude 148°00'W longitude 12°30'S latitude 158°00'W longitude	Earth parking orbit insertion to TLI Entry minus 6 hours to landing

TABLE 5 (Continued)

Nominal Mission Phase Aircraft

Designation	Location	Mission coverage
3 HC-130	Ascension	Earth parking orbit insertion to entry minus 24 hours.
2 HC-130	Howard APB, Canal Zone	Earth parking orbit insertion to entry minus 24 hours.
1 HC-130	Mauritius Island	Earth parking orbit insertion to entry minus 24 hours.
2 HC-130	Anderson APB, Guam	Earth parking orbit insertion to entry minus 24 hours.
2 HC-130	Hickam APB, Hawaii	Earth parking orbit insertion to entry minus 24 hours.
4 HC-130	Pago Pago, Samoa	Support primary landing area.

Logistic Aircraft

1 Helicopter (2 flights)	PRS to Samoa	Transport flight crew and NASA personnel
1 C-141	Cape Kennedy or Long Beach Municipal Airport to CM deactivation site.	Transport deactivation equipment and personnel
	Then, Samoa to Ellington APB, Texas	Transport flight crew and NASA personnel
1 C-133B	Hickam APB, Hawaii to Long Beach Municipal Air-Dort	Return CM to North American Aviation. Inc.

EARTH PARKING ORBIT PHASE

Earth Orbital Secondary Landing Areas

An earth orbital secondary landing area is where the probability of landing after insertion and prior to TLI is sufficiently high to require secondary recovery ship support. For Apollo 10 these areas are 210 by 80 nautical mile ellipses centered on the target point and oriented with the major axis along the entry ground track. Planned recovery forces will have a capability to provide a maximum access time of 6 hours to any point in the area, and a maximum crew and CM retrieval time of 40 hours to any point in the area.

Earth Orbital Contingency Landing Area

The contingency landing area is that area where the probability of landing is very low and requires only land-based aircraft support. The contingency landing area during the earth orbital phase of Apollo 10 is all the area in a band around the earth between 30°N and 34°S that lies outside the earth orbital secondary landing areas.

The recovery forces will have a capability to achieve a maximum access time of 18 hours in the major portion of the area. It is accepted that portions of the area lie outside the 18-hour capability of the aircraft. In these portions of the area the probability of landing is extremely small and the access time requirement is as soon as possible with no maximum limit defined.

TRANSLUNAR INJECTION TO END OF MISSION

Deep Space Secondary Landing Areas

A deep space secondary landing area is an area where the probability of landing after TLI is sufficiently high to require secondary recovery ship support. For the Apollo 10 Mission, these areas are defined as the areas where landing would occur following (1) translunar coast aborts targeted to the Mid-Pacific line and (2) any abort after TLI targeted to the Atlantic Ocean line (Figure 20).

The recovery forces will have a capability to achieve a maximum access time of 10 hours to any point in the area and a maximum crew and CM retrieval time of 26 hours to any point in the area.

RECOVERY LINES

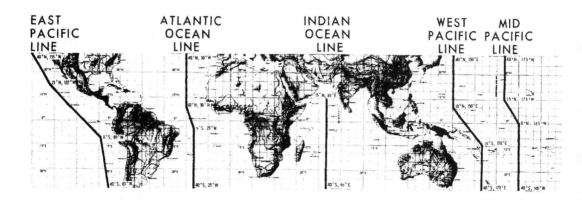

Fig 20.

Deep Space Contingency Landing Area

The contingency landing area for the deep space phase of the mission is associated with very low probabilities of landing and requires land-based recovery aircraft support only.

Recovery forces will provide a capability to achieve a maximum access time of 18 hours to any point on the West Pacific, East Pacific, and Indian Ocean recovery lines after TLI. Certain portions of the recovery lines and some of the area between lines lie outside of the 18-hour capability of aircraft. Since the probability of landing in these areas is extremely small this is acceptable and no maximum access time is defined for these landings. The aircraft required for the West Pacific, East Pacific, and Indian Ocean recovery lines, coupled with the aircraft required for the secondary landing areas previously described, will adequately support these portions of the area.

END OF MISSION

The primary landing area is that area where the probability of landing is sufficiently high to warrant a requirement for primary recovery ship support. For Apollo 10, the primary landing area (Figure 21) is where the spacecraft will land following circumlunar or lunar orbital trajectories that are targeted to the Mid-Pacific recovery line. End of mission landing points for various Apollo 10 launch dates is shown in Figure 22.

Recovery forces will provide a capability to achieve a maximum access time of 2 hours to any point in the area, a maximum crew retrieval time of 16 hours to any point in the area, a maximum CM retrieval time of 24 hours to any point in the area, and a helicopter from which photographers may take pictures of the CM as soon as possible after landing.

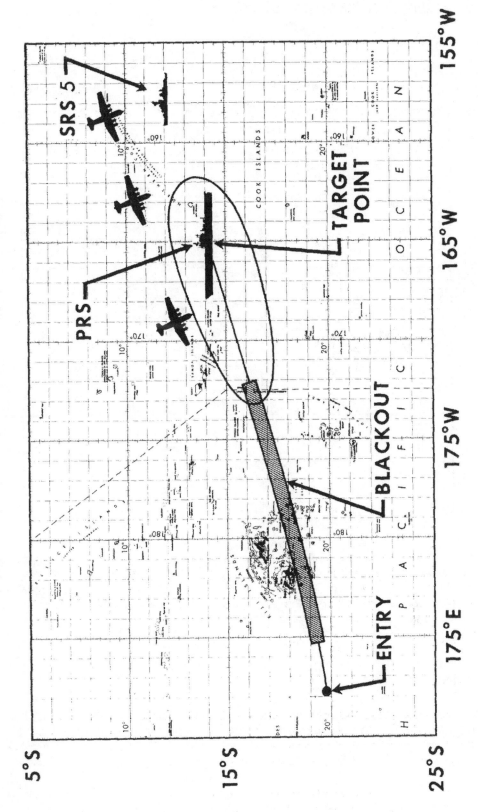

PRIMARY LANDING AREA

Fig 21.

END-OF-MISSION LANDING POINTS

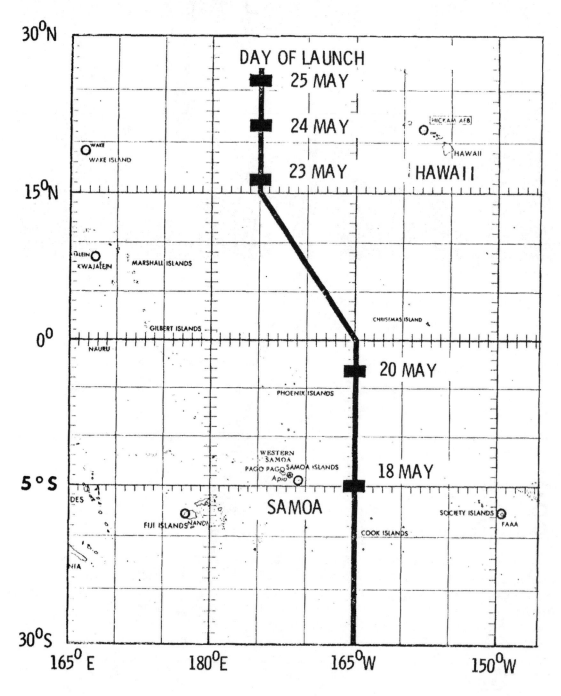

Fig 22.

APOLLO 10 PRIME CREW   Fig 23.

THOMAS P. STAFFORD   JOHN W. YOUNG   EUGENE A. CERNAN

FLIGHT CREW

FLIGHT CREW ASSIGNMENTS

Prime Crew (Figure 23)

Commander (CDR) - T. P. Stafford (Colonel, USAF) Command Module Pilot (CMP) - J. W. Young (Commander, USN) Lunar Module Pilot (LMP) - E. A. Cernan (Commander, USN)

Backup Crew (Figure 24)

Commander (CDR) - L. G. Cooper (Colonel, USAF) Command Module Pilot (CMP) - D. F. Eisele (Lt. Colonel, USAF) Lunar Module Pilot (LMP) - E. D. Mitchell (Commander, USN)

If necessary, the backup crew can be substituted for the prime crew up to about two weeks prior to an Apollo launch. During this period, the flight hardware and software, ground hardware and software, flight crew and ground crews work as an integrated team to perform ground simulations and other tests of the upcoming mission. It is necessary that the flight crew that will conduct the mission take part in these activities, which are not repeated for the benefit of the backup crew. To do so would add an additional costly two-week period to the prelaunch schedule, which for a lunar mission would require rescheduling for the next lunar window.

PRIME CREW BIOGRAPHICAL DATA

Commander (CDR)

NAME: Thomas P. Stafford (Colonel, USAF)

BIRTHPLACE AND DATE: Weatherford, Oklahoma; 17 September 1930.

PHYSICAL DESCRIPTION: Black hair; blue eyes; height: 6 ft; weight: 175 lbs.

EDUCATION: Graduated from Weatherford High School, Weatherford, Oklahoma; received a Bachelor of Science degree from the United States Naval Academy in 1952; recipient of an Honorary Doctorate of Science from Oklahoma City University in 1967.

ORGANIZATIONS: Member of the Society of Experimental Test Pilots.

SPECIAL HONORS- Awarded two NASA Exceptional Service Medals and the Air Force Astronaut Wings; the Distinguished Flying Cross; the AIAA Astronautics Award; and co-recipient of the 1966 Harmon International Aviation Trophy.

EXPERIENCE: Stafford, commissioned in the United States Air Force upon graduation from Annapolis, flew fighter interceptor aircraft in the United States and Germany after finishing his flight training. He later attended the USAF Experimental Flight Test School at Edwards Air Force Base, California and then served as Chief of the Performance Branch in the USAF Aerospace Research Pilot School there. He is co-author of the Pilots' Handbook for Performance Flight Testing and the Aerodynamics Handbook for Flight Testing.

CURRENT ASSIGNMENT: Colonel Stafford was selected as an astronaut by NASA in September 1962.

On 15 December 1965, he and Command Pilot Walter M. Schirra were launched into space on the Gemini 6 mission and participated in the first successful rendezvous of two manned maneuverable spacecraft by joining the already orbiting Gemini 7 crew. On his second flight he was Command Pilot of the Gemini 9, a 3-day mission which began on 3 June 1966. The spacecraft attained a circular orbit of 161 statute miles; the crew performed three different types of rendezvous with the previously launched augmented Target Docking Adapter; and Pilot Eugene Cernan logged 2 hours 10 minutes outside the spacecraft in extravehicular activity.

Command Module Pilot (CMP)

NAME: John W. Young (Commander, USN)

BIRTHPLACE AND DATE: San Francisco, California, 24 September 1930.

PHYSICAL DESCRIPTION: Brown hair; green eyes; height: 5 ft 9 in; weight: 165 lbs.

EDUCATION: Graduated from Orlando High School, Orlando, Florida; received a Bachelor of Science degree in Aeronautical Engineering from the Georgia Institute of Technology in 1952.

ORGANIZATIONS: Member of the American Institute of Aeronautics and Astronautics and the Society of Experimental Test Pilots.

SPECIAL HONORS: Awarded two NASA Exceptional Service Medals, the Navy Astronaut Wings, and three Distinguished Flying Crosses.

EXPERIENCE: Upon graduation from Georgia Institute of Technology, Young entered the US Navy in 1952. He was a test pilot at the Naval Air Test Center from 1959 to 1962 and set world time-to-climb records to 3000 and 25,000 meter altitudes in the F4B in 1962. Prior to his assignment to NASA he was Maintenance

Officer of Al 1 -Weather- Fighter Squadron 143 at the Naval Air Station, Miramar, California.

CURRENT ASSIGNMENT: Commander Young was selected as an astronaut by NASA in September 1962.

He served as Pilot on the first manned Gemini flight on 3 March 1965, during which the crew accomplished the first manned spacecraft orbital trajectory modifications and lifting reentry.

On 18 July 1966, Young was Command Pilot for the Gemini 10 mission and, with Michael Collins as Pilot, effected a successful rendezvous and docking with the Agena target vehicle.

<u>Lunar Module Pilot (LMP)</u>

NAME: Eugene A. Cernan (Commander, USN)

BIRTHPLACE AND DATE: Chicago, Illinois, 14 March 1934.

PHYSICAL DESCRIPTION: Brown hair; blue eyes; height: 6 ft; weight: 170 lbs.

EDUCATION: Graduated from Proviso Township High School in Maywood, Illinois; received a Bachelor of Science degree in Electrical Engineering from Purdue University and a Master of Science degree in Aeronautical Engineering from the US Naval Postgraduate School.

ORGANIZATIONS: Member of Tau Beta Pi, national engineering society; Sigma Xi, national science research society; and Phi Gamma Delta, national social fraternity.

SPECIAL HONORS: Awarded two Distinguished Flying Crosses, two NASA Exceptional Service Medals, and the Navy Astronaut Wings; recipient of Princeton's Distinguished Alumnus Award for 1965, the US Jaycee's 10 Outstanding Young Men Award in 1965, and the American Astronautical Society Flight Achievement Award for 1966.

EXPERIENCE: Cernan received his commission through the naval ROTC program at Purdue and entered flight training upon his graduation. Prior to attending the Naval Postgraduate School, he was assigned to Attack Squadrons 125 and 113 at the Miramar, California, Naval Air Station.

CURRENT ASSIGNMENT: Commander Cernan was one of the third group of astronauts selected by NASA in October 1963.

On the Gemini 9 mission on 3 June 1966, he was Pilot with Command Pilot Tom Stafford and participated in three different techniques to effect rendezvous with the previously launched Augmented Target Docking Adapter. During the 3-day flight, he logged 2 hours 10 minutes outside the spacecraft in extravehicular activity.

He also served as backup pilot for Gemini 12.

Fig 24.   A P O L L O   1 0   B A C K - U P   C R E W

L. GORDON COOPER   DONN F. EISELE   EDGAR D. MITCHELL

BACKUP CREW BIOGRAPHICAL DATA

Commander (CDR)

NAME: L. Gordon Cooper (Colonel, USAF)

BIRTHPLACE AND DATE: Shawnee, Oklahoma, 5 March 1927.

PHYSICAL DESCRIPTION: Brown hair; blue eyes; height; 5 ft 8 in; weight: 150 lbs.

EDUCATION: Attended primary and secondary schools in Shawnee, Oklahoma and Murray, Kentucky; received a Bachelor of Science degree in Aeronautical Engineering from the Air Force Institute of Technology (AFIT) in 1956; recipient of an Honorary Doctorate of Science from Oklahoma City University in 1967.

ORGANIZATIONS: Member of American Institute of Aeronautics and Astronautics, the Society of Experimental Test Pilots, the American Astronautical Society, the Blue Lodge Masons, the Scottish Rite Masons, the York Rite Masons, the Shrine, the Jesters, the International Rotary Club, and the Confederate Air Force.

SPECIAL HONORS: Awarded the NASA Distinguished Service Medal, the NASA Exceptional Service Medal, Air Force Command Astronauts Wings, Distinguished Flying Cross with cluster, the Air Force Command Missileman's Badge, the Scottish Rite 330, and the York Rite Knight of the Purple Cross.

EXPERIENCE: Colonel Cooper received an Army Commission after completing three years of schooling at the University of Hawaii. He transferred his commission to the Air Force and was placed on active duty in 1949. He then flew F-84's and F-86's for four years with the 86th Fighter Bomber Group in Munich, Germany. He returned to the United States and, after two years of study at AFIT, reported to the Air Force Experimental Flight Test School at Edwards Air Force Base, California. Upon graduation in 1957, he was assigned as an aeronautical engineering and test pilot in the Performance Engineering Branch of the Flight Test Division at Edwards.

CURRENT ASSIGNMENT: Colonel Cooper was selected as a Mercury astronaut in April 1959.

On 15-16 May 1963, he commanded the "Faith 7" spacecraft on a 22-orbit mission which concluded the operational phase of Project Mercury.

He served as Command Pilot of the 8-day, 120-revolution Gemini 5 mission which began on 21 August 1965 and, with Pilot Charles Conrad, established a new space endurance record. Cooper also become the first man to make a second orbital flight.

Command Module Pilot (CMP)

NAME: Donn F. Eisele (Lt. Colonel, USAF)

BIRTHPLACE AND DATE: Columbus, Ohio, 23 June 1930.

PHYSICAL DESCRIPTION: Brown hair; blue eyes; height: 5 ft 9 in; weight: 150 lbs.

EDUCATION: Graduated from West High School, Columbus, Ohio; received a Bachelor of Science degree from the United States Naval Academy in 1952 and a Master of Science degree in Astronautics in 1960 from the Air Force Institute of Technology, Wright-Patterson Air Force Base, Ohio.

ORGANIZATIONS: Member of Tau Beta Pi (national engineering society).

EXPERIENCE: Eisele chose a career in the Air Force after graduating from the US Naval Academy. He is also a graduate of the Air Force Aerospace Research Pilot School at Edwards Air Force Base, California.

He was a project engineer and experimental test pilot at the Air Force Special Weapons Center at Kirkland Air Force Base, New Mexico.

CURRENT ASSIGNMENT: Lt. Colonel Eisele was one of the third group of astronauts selected by NASA in October 1963. He served as Command Module Pilot on Apollo 7.

Lunar Module Pilot (LMP)

NAME: Edgar Dean Mitchell (Commander, USN)

BIRTHPLACE AND DATE: Hereford, Texas, 17 September 1930.

PHYSICAL DESCRIPTION: Brown hair; green eyes; height: 5 ft 11 in; weight: 180 lbs.

EDUCATION: Graduated from Artesia High School, Artesia, New Mexico; received a Bachelor of Science degree in Industrial Management from the Carnegie Institute of Technology in 1952, a Bachelor of Science degree in Aeronautical Engineering from the US Naval Postgraduate School in 1961, and a Doctor of Science degree in Aeronautics/Astronautics from the Massachusetts Institute of Technology in 1964.

ORGANIZATIONS: Member of the American Institute of Aeronautics and Astronautics; Sigma Xi; and Sigma Gamma Tau.

EXPERIENCE: Commander Mitchell entered the Navy in 1952 and completed his basic training at the San Diego Recruit Depot. In May 1953, after completing instruction at the Officer's Candidate School at Newport, Rhode Island, he was commissioned as an Ensign. He completed his flight training in July 1954 at Hutchinson, Kansas, and was assigned to Patrol Squadron 29 deployed to Okinawa.

From 1957 to 1958, he flew A3 aircraft while assigned to Heavy Attack Squadron Two deployed aboard the USS BON HOMME RICHARD and USS TICONDEROGA; and he was a Research Project Pilot with Air

Development Squadron Five until 1959. His assignment from 1964 to 1965 was as Chief, Project Management Division, of the Navy Field Office for Manned Orbiting Laboratory.

CURRENT ASSIGNMENT: Commander Mitchell was in the group selected for astronaut training in April 1966.

MISSION MANAGEMENT RESPONSIBILITY

Title	Name	Organization
Director, Apollo Program	Lt. Gen. Sam C. Phillips	NASA/OMSF
Director, Mission Operations	Maj. Gen. John D. Stevenson (Ret)	NASA/OMSF
Saturn V Vehicle Prog. Mgr.	Mr. Lee B. James NASA/MSFC	
Apollo Spacecraft Prog. Mgr.	Mr. George M. Low	NASA/MSC
Apollo Prog. Manager KSC	R. Adm. Roderick O. Middleton	NASA/KSC
Mission Director	Mr. George H. Hoge	NASA/OMSF
Assistant Mission Director	Capt. Chester M. Lee (Ret)	NASA/OMSF
Assistant Mission Director	Col. Thomas H. McMullen	NASA/OMSF
Director of Launch Operations	Mr. Rocco Petrone	NASA/KSC
Director of Flight Operations	Mr. Christopher C. Kraft	NASA/MSC
Launch Operations Manager	Mr. Paul C. Donnelly	NASA/KSC
Flight Directors	Mr. Glynn S. Lunney	NASA/MSC
	Mr. M. P. Frank	
	Mr. Gerald D. Griffin	
Spacecraft Commander (Prime)	Col. Thomas P. Stafford	NASA/MSC
Spacecraft Commander (Backup)	Col. L. G. Cooper	NASA/MSC

PROGRAM MANAGEMENT

NASA HEADQUARTERS
Office of Manned Space Flight
Manned Spacecraft Center
Marshall Space Flight Center
Kennedy Space Center

LAUNCH VEHICLE	SPACECRAFT	TRACKING AND DATA ACQUISITION
Marshall Space Flight Center The Boeing Co. (S-IC)	Manned Spacecraft Center	Kennedy Space Center
North American Rockwell Corp. (S-II)	North American Rockwell (LES, CSM, SLA)	Goddard Space Flight Center
McDonnell Douglas Corp. (S-IVB)	Grumman Aircraft Engineering Corp. (LM)	Department of Defense
IBM Corp. (IU)		MSFN

ABBREVIATIONS

AGS	Abort Guidance System
AOL	Atlantic Ocean Line
AOS	Acquisition of Signal
APS	Ascent Propulsion System (LM)
APS	Auxiliary Propulsion System (S-IVB)
AS	Ascent Stage
BPC	Boost Protection Cover
CDH	Constant Delta Height
CDR	Commander
CES	Control Electronics System
CM	Command Module
CMP	Command Module Pilot
COI	Contingency Orbit Insertion
CSI	Concentric Sequence Initiation
CSM	Command/Service Module
DOI	Descent Orbit Insertion
DPS	Descent Propulsion System
DS	Descent Stage
DTO	Detailed Test Objective
EDS	Environmental Control System
EDS	Emergency Detection System
EDT	Eastern Daylight Time
EI	Entry Interface
EPS	Electrical Power System
EPO	Earth Parking Orbit
EVA	Extravehicular Activity
GET	Ground Elapsed Time
GHe	Gaseous Helium
GN&CS	Guidance, Navigation, and Control System
GOX	Gaseous Oxygen
IMU	Inertial Measurement Unit
IS	Instrumentation System
IU	Instrument Unit
IVT	Intervehicular Transfer
KSC	Kennedy Space Center
LC	Launch Complex
LCC	Launch Control Center
LES	Launch Escape System
LET	Launch Escape Tower
LH2	Liquid Hydrogen
LM	Lunar Module
LMP	Lunar Module Pilot
LOI	Lunar Orbit Insertion
LOX	Lunar Orbit Rendezvous
LOS	Loss of Signal
LOX	Liquid Oxygen
LPO	Lunar Parking Orbit
LV	Launch Vehicle
MCC	Midcourse Correction
MCC	Mission Control Center
MOCR	mission operations Control Room
MOR	Mission operation Report
MPL	Mid-Pacific Line

MSFN	Manned Space Flight Network
MSS	Mobile Service Structure
NM	Nautical Mile
PC	Plane Change
PDI	Powered Descent Initiation
PGNCS	Primary Guidance, Navigation, and Control System
PRS	Primary Recovery Ship
PTC	Passive Thermal Control
PTP	Preferred Target Point
RCS	Reaction Control System
RR	Rendezvous Radar
SAR	Search and Rescue.
SC	Spacecraft
SCS	Stabilization and Control System
SEA	Sun Elevation Angle
SEQ	Sequential System
SHe	Supercritical Helium
SLA	Spacecraft LM Adapter
SM	Service Module
SPS	Service Propulsion System
SRS	Secondary Recovery Ship
SV	Space Vehicle
TB	Time Base
T,D&E	Transposition, Docking, and Ejection
T/C	Telecommunications
TEC	Transearth Coast
TEI	Transearth Insertion
TLC	Translunar Coast
TLI	Translunar Injection
TPF	Terminal Phase Finalization
TPI	Terminal Phase Initiation
T-time	Countdown time (referenced to lift-off time)
TV	Television
VAB	Vehicle Assembly Building
VHF	Very High Frequency

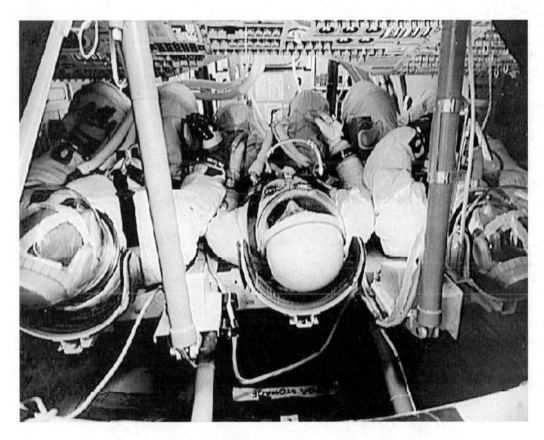

The Apollo 10 crew during simulator training before the flight.

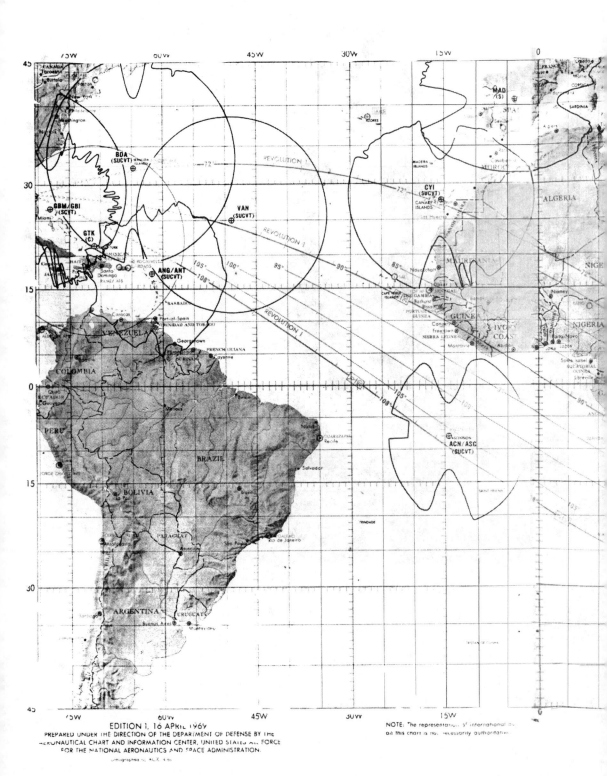

EDITION 1, 16 APRIL 1969
PREPARED UNDER THE DIRECTION OF THE DEPARTMENT OF DEFENSE BY THE
AERONAUTICAL CHART AND INFORMATION CENTER, UNITED STATES AIR FORCE
FOR THE NATIONAL AERONAUTICS AND SPACE ADMINISTRATION.

NOTE: The representation of international
on this chart is not necessarily authoritative.

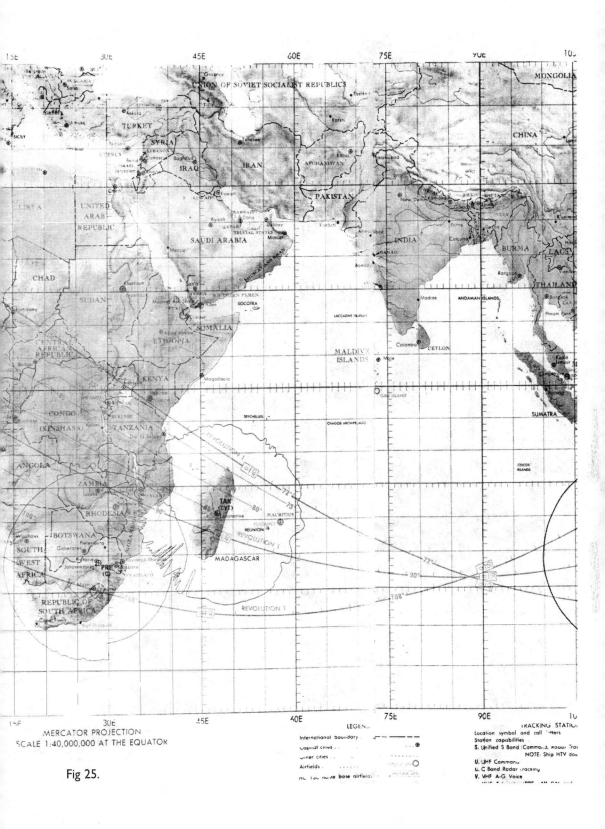

MERCATOR PROJECTION
SCALE 1:40,000,000 AT THE EQUATOR

Fig 25.

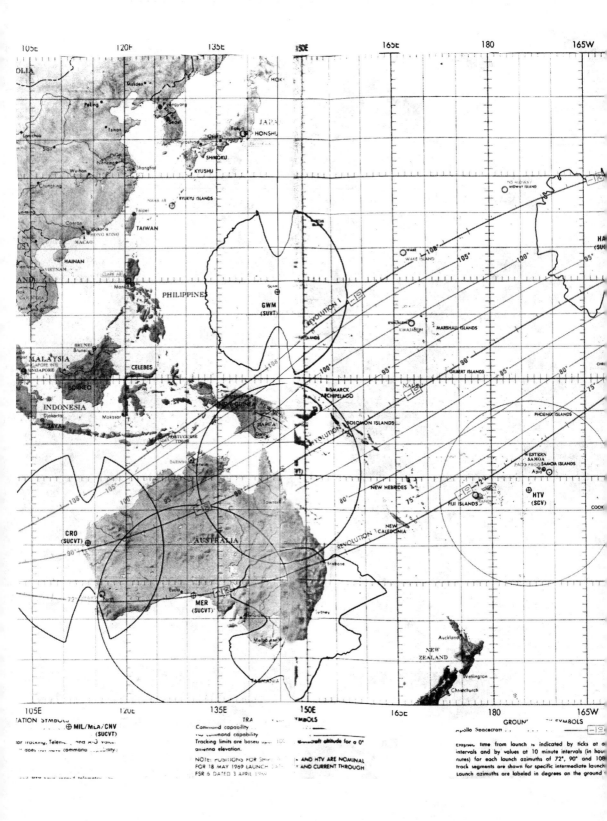

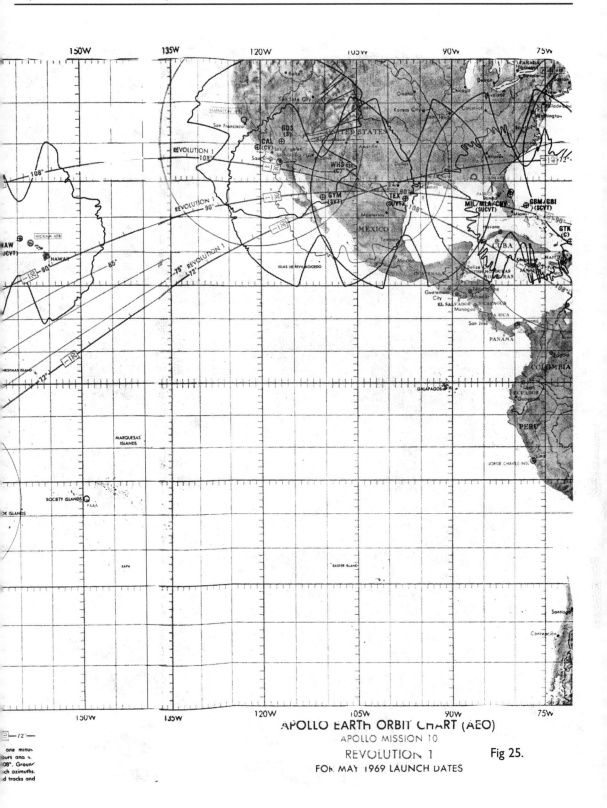

APOLLO EARTH ORBIT CHART (AEO)
APOLLO MISSION 10
REVOLUTION 1   Fig 25.
FOR MAY 1969 LAUNCH DATES

NATIONAL AERONAUTICS
AND SPACE ADMINISTRATION
WASHINGTON, D.C. 20546

POST LAUNCH
MISSION OPERATION
REPORT

No. M-932-69-10

26 May 1969

TO: A/Administrator

FROM: MA/Apollo Program Director

SUBJECT: Apollo 10 Mission (AS-505) Post Launch Mission Operation Report No. 1

The Apollo 10 mission was successfully launched from the Kennedy Space Center on Sunday, 18 May 1969 and was completed as planned, with recovery of the spacecraft and crew in the Pacific Ocean recovery area on Monday, 26 May 1969. Initial review of the flight indicates that all mission objectives were attained.

Attached is the Mission Director's Summary Report for Apollo 10 which is hereby submitted as Post Launch Mission Operation Report No. 1. Following further detailed analysis of data, crew briefing and other technical reviews, significant new information and OMSF evaluation of Apollo 10 primary mission objectives will be submitted in Post Launch Mission Operation Report No. 2.

Sam C. Phillips
Lt. General, USAF
Apollo Program Director

NATIONAL AERONAUTICS AND SPACE ADMINISTRATION WASHINGTON, D.C. 20546

IN REPLY REFER TO: MAO May 26, 1969

TO: Distribution

FROM: MA/Apollo Mission Director

SUBJECT: Mission Director's Summary Report, Apollo 10

INTRODUCTION

The Apollo 10 mission was planned as a manned lunar mission development flight to demonstrate crew/space vehicle/mission support facilities performance during a manned lunar mission with the Command/Service Module (CSM) and Lunar Module (LM), and to evaluate LM performance in the cislunar and lunar environment. Flight crew members were: Commander, Col. T. P. Stafford; Command Module Pilot, Cdr. J. W. Young Lunar Module Pilot., Cdr. E. A. Cernan. Initial review of the flight indicates that all mission objectives were attained (See Table 1).

PRELAUNCH

The Apollo 10 countdown was accomplished with no unscheduled holds. Prior to launch, the Command Module Reaction Control System A helium manifold pressure decayed slightly. A suspect transducer fitting was found to be finger-tight and was subsequently retorqued. Difficulty was encountered in wetting the sintered metal wicks in the Command Module suit loop water separator, a component of the Environmental Control System. The procedure was successfully accomplished on the third attempt to service.

FIRST PERIOD

Major activity in this period included space vehicle launch, insertion into earth orbit and translunar injection of the S-IVB/ instrument unit (IU) /CSM / LM transposition, docking and ejection of the CSM/ LM and S-IVB propellant dump injecting the S-IVB into solar orbit.

The Apollo 10 space vehicle was successfully launched on time from Kennedy Space Center, Florida at 12:49 p.m. EDT on 18 May 1969. This was the fifth successive successful launch on-time of a Saturn V. All launch vehicle stages performed satisfactorily, inserting the S-IVB/IU/CSM/LM combination into a nominal earth parking orbit of 102.6 by 99.6 nautical miles (NM) after 11minutes 53 seconds of powered flight. Pre-TLI (translunar injection) checkout was conducted as planned and the S-IVB burn was initiated at 2:33:27 (hr:min:sec) ground elapsed time (GET). The TLI burn lasted 5 minutes, 43 seconds with all systems operating satisfactorily and all end conditions being nominal for the translunar coast on a free return circumlunar trajectory.

At about 3:03 (hr:min) GET, the SM was separated normally from the rest of the orbital vehicle consisting of the LM, Spacecraft LM Adapter (SLA), IU and S-IVB Stage. SLA panel deployment was normal. CSM transposition and docking were completed by approximately 3:17 GET. Excellent quality color television (TV) coverage of the docking sequences was transmitted to the Goldstone tracking station and was seen on worldwide commercial television. Ejection of the CSM/LM from the S-IVB was successfully accomplished at about 5:56 GET and a 2.5-second Service Propulsion System (SPS) evasive maneuver was performed as planned at 4:39 GET.

All launch vehicle safing activities were performed as scheduled. S-IVB liquid oxygen and liquid hydrogen lead temperature measurement experiments were conducted satisfactorily. Subsequent propellant dump was successful and provided sufficient impulse to the S-IVB/IU for a "slingshot" maneuver to earth escape velocity. Therefore, augmentation of this impulse by the S-IVB Auxiliary Propulsion System ullage engine burn was terminated by ground command immediately after ignition. S-IVB/IU closest approach to the moon was 1752 nm at 78:54 GET (19:43 EDT, May 21).

SECOND PERIOD

Major activities during the second period were a midcourse correction two lunar orbit insertion burns and initial LM activation.

Midcourse correction maneuver number 1 (MCC-1) originally planned as a 47 foot-per-second (fps) SPS maneuver, was not conducted at 11:30 GET, since without this maneuver, the correction requirement at the next planned time, 26:39 GET, was for only 48.8 fps. The extended tracking time established a high probability of not requiring any additional midcourse corrections during translunar coast.

Midcourse correction maneuver number two (MCC-2) was performed at 26:32:56 GET by a 6.7-second firing of the SPS resulting in a velocity change of 48.9 fps (48.7 fps planned). All parameters appeared nominal and the resulting pericynthion was 60.9 nm. Consequently, midcourse correction maneuvers numbers 3 and 4 were not required. Five color TV transmissions totaling 72 minutes were made during translunar coast. Views of the receding earth and of the spacecraft interior were shown. Picture quality was excellent.

The spacecraft crossed into the moon's sphere of influence at 61:50:50 GET (02:39 EDT, May 21). At that time, the distance from the spacecraft to earth was 190,535 nm and its distance from the moon was 33,820 nm. The velocity was 3120 fps relative to earth and 3795 fps relative to the moon.

The lunar orbit insertion maneuver (LOI-1) was planned in real time for 75:55:53 GET and was accomplished on schedule. The SPS Engine burned for 356 seconds slowing the CSM/LM down from a
velocity of 8222 fps to a velocity of 5472 fps and resulting in an initial orbit of 170.4 by 59.6 nm. This compares very well with the prelaunch planned orbit of 170 by 60 nm, and the realtime planned orbit of 170.7 by 59.7 nm. The SPS burn data appeared to be nominal with fuel tank pressure and oxidizer interface pressure slightly on the high side of nominal, but well within expected tolerances. Spacecraft weight at initiation of the burn was 93,281 pounds and at termination of the burn, spacecraft weight in lunar orbit was 69,493 pounds. Propellant expended for the burn was 23,788 pounds.

The lunar orbit circularization maneuver, LOI-2, was planned in time for 80:25:07 GET and was also accomplished on schedule. The SPS engine burned for 13.9 seconds producing a differential velocity of 138.4 fps and resulting in an initial orbit of 61.5 by 58.9 nm. This compares well with the prelaunch planned orbit of 60 by 60 nm and the real-time planned orbit of 60.1 by 60.1 nm. All SPS parameters were nominal.

A 29-minute scheduled color television transmission of the lunar surface was conducted 80:45 GET (21:34 EDT, 21 May). Picture quality of lunar scenes was excellent.

Lunar landmark tracking on two targets was accomplished and indications are that these landmarks were well spaced and of good quality.

The Lunar Module Pilot transferred to the LM at about 81:55 GET for about two hours of scheduled "housekeeping" activities and some LM communications tests. The tests were terminated after the LM relay communications tests due to time limitations. Results of completed tests were excellent and those tests remaining were conducted at a later time in the mission.

THIRD PERIOD

Major activities in this period were the LM descent to within 50,000 feet of the lunar surface and subsequent rendezvous with the orbiting CSM.

The Commander and Lunar Module Pilot entered the LM at 95:02 GET and performed the preplanned checks of all systems. The rendezvous exercise was begun on time with undocking at approximately 98:22 GET. The Service Module Reaction Control System (SM RCS) was used to separate the CSM to about 50 feet from the LM. Subsequently the landing gear was deployed. Stationkeeping was initiated at this point while the Command Module Pilot in the CSM visually inspected the LM. The SM RCS was then used to perform the separation maneuver directed radially downward toward the moon's center. This maneuver provided a LM/CSM separation at descent orbit insertion (DOI) of about two nm. The DOI was performed by a LM Descent Propulsion System (DPS) burn (horizontal, retrograde), such that the resulting pericynthion (lowest point in orbit) occurred about 15° prior to lunar landing site number 2. The lowest altitude above the moon's surface achieved by the LM was 8.4 nm. Numerous photographs of the lunar surface were taken. Some camera malfunctions were reported and although some communications difficulties were experienced, the crew provided a continuous commentary of their observations. The LM landing radar test was executed during the low altitude pass over the surface. Early data indicates initial acquisition occurred at a height of 65,000 feet. Indicated pericynthion altitude as measured by the landing radar in the fly-by was 47,000 feet.

The second LM maneuver, the DPS phasing burn, was accomplished on time and established (as planned) at the resulting LM pericynthion (about one revolution later), a CSM lead angle equivalent to that which occurs at nominal powered ascent cutoff for the lunar landing mission. The apocynthion (orbital high point) altitude of the phasing orbit was 190.0 nm.

About ten minutes prior to pericynthion, the LM Descent Stage was jettisoned. The LM Reaction Control System (LM RCS) separation maneuver at staging was accomplished using the Abort Guidance System (AGS) as prescribed in premission plans. Inadvertently, the control mode was left in AUTO rather than the required ALTITUDE HOLD mode. In AUTO, the AGS drove the LM to acquire the CSM which was not in accordance with the planned attitude timeline. The Commander took over manual control to reestablish the proper attitude. Then at pericynthion, the insertion maneuver was performed on time using the LM Ascent Propulsion System (APS). This burn established the equivalent of the standard LM insertion orbit (45 x 11.2 nm) of a lunar landing mission.

The LM coasted from insertion in a 45 nm by 11.2 nm elliptical orbit for about an hour. Concentric sequence initiation (CSI) was initiated at apocynthion. A small constant delta height (CDH) maneuver was required (as expected preflight) to null out minor dispersions. The terminal maneuver occurred at about the midpoint of darkness, and braking during the terminal phase finalization (TPF) was performed manually as planned.

The rendezvous was highly successful and all parameters were very close to nominal. CSM-active docking at 106.33 GET was accomplished smoothly and expeditiously.

Once docked to the CSM, the two LM crewmen transferred with the exposed film packets and the LM Hasselblad camera to the CSM. The LM Maurer sequence camera and primary lithium hydroxide canister (both of which incurred inflight problems) were also transferred in order that these items could be inspected post flight. The CSM was separated from the LM at 108:43:30 GET using the SM RCS.

FOURTH PERIOD

Major activity in the fourth period included the LM APS burn to depletion, extensive landmark tracking, photography, TV and the transearth injection burn.

About one revolution after docking, the LM APS burn to depletion was commanded by the Manned Space Flight Network (MSFN), as planned, utilizing the LM Ascent Engine Arming Assembly. This burn was targeted to place the LM in a solar orbit. LM/MSFN communications were maintained until LM ascent stage battery depletion at about 120 hours GET. The ascent batteries lasted about 12 hours after LM jettison.

During the remaining lunar orbital period of operation, 18 landmark sightings and extensive stereo strip and oblique photographs were taken. Two scheduled TV periods were deleted because of crew fatigue. The crew visually acquired the LM descent stage on several occasions. At 137:36:28 GET, the SPS injected the CSM into a transearth trajectory after a total time in lunar orbit of about 61.5 hours (31 orbits). The TEI burn was targeted for a transearth return time of about 53 hours.

FIFTH PERIOD

Major activities during this period included star-lunar landmark sightings, live color television transmissions, star-earth horizon navigation sightings, CSM S-band high gain antenna reflectivity test and MCC-7.

This period commenced with a live television transmission through the Honeysuckle tracking station and Intelsat III communications satellite beginning shortly after TEI at about 137:51 GET. Pictures of the moon as seen from the receding spacecraft were spectacular. Focus at all "zoom" lens settings was excellent and color fidelity of lunar surface scenes agreed very well with crew descriptions.

Another color television transmission was received at 139:27 GET.

Following a sleep period, star lunar landmark navigation sightings were taken at 151:00 GET.

The accuracy of the transearth injection (TEI) maneuver was such that the first transearth midcourse correction (MCC-5) originally scheduled for 152:00 GET was not necessary. The Atlantic Ocean contingency recovery forces were released from mission support at 153:00 GET

The waste water dump conducted at 153:50 was oriented to reduce the probability of midcourse corrections. Checkout of the Entry Monitor System was accomplished at 154:35 GET to ensure its readiness for the entry phase.

A ten-minute color television broadcast was made at approximately 147:23 GET. Earth scenes were shown for about three minutes with moon scenes throughout the remainder of the broadcast. A twenty nine minute broadcast of the moon, earth and spacecraft interior was received at 152:29 GET.

A number of star-earth horizon navigation sightings were taken. The CSM S-band high gain reflectivity test was conducted at 168:00 GET. An unscheduled live color television transmission of the earth and the command module interior was received at 173:27 GET. The second transearth midcourse correction (MCC-6) originally scheduled for 176:50 GET was not necessary.

SIXTH PERIOD

Major activities during this period included MCC-7 live color television, reentry and recovery.

The crew was awakened at 185:00 GET and initiated reentry preparations. The final live color television transmission was received at 186:50 GET. MCC-7 was performed at 188:49 GET. Entry interface was reached at 191:48:54 GET with splashdown in the mid Pacific, approximately 165°W and 15°S.

Weather in the prime recovery area was excellent; visibility, 10 miles, wave height 4 feet, cloud cover 1800 feet scattered, winds less than 12 knots.

RESCUE 1 reported visual contact of the spacecraft at 191:52. AIRBOSS 1 visually acquired the spacecraft one minute later. Voice communications with the crew were reported by the USS PRINCETON at 191:53:42 GET. Drogue and main parachutes deployed normally. Splashdown occurred about 14 minutes after entry interface at 192:03:23 GET, approximately 3 nm from the prime recovery ship, USS PRINCETON. The Command Module remained in the Stable 1 position, and the crew reported that they were in good condition. The crew was picked up by a recovery helicopter and was safe aboard the ship at 13:31 EDT May 26, 1969, to end a fantastic mission.

SYSTEMS PERFORMANCE

All launch vehicle systems performed satisfactorily throughout their expected lifetimes. All spacecraft systems continued to function satisfactorily throughout the mission with the exception of fuel cell No. 1.

At 120:47 GET, fuel cell No. I experienced an electrical failure in the cooling pump circuit and was isolated from the main bus. It was placed back on the bus for the transearth injection maneuver and with close monitoring of temperature limits, provided satisfactory voltages and currents. Subsequently an effective purging cycle was initiated and the fuel cell was taken off circuit for the remainder of the mission. It remained available for load sharing, if required.

A number of other minor discrepancies occurred which were primarily procedural, and were corrected in flight with no mission impact, or involved instrumentation errors on quantities which could be checked by other means. Temperature and consumables usage rates remained generally within normal limits throughout the Mission. Complete analysis of systems performance will be reported in subsequent MSF Center engineering reports.

FLIGHT CREW PERFORMANCE

Flight crew performance was outstanding. All three crew members remained in excellent health throughout the mission. Their prevailing good spirits were continually evident as they took time from their busy schedule to share their voyage with the world via 19 color television transmissions totaling almost six hours.

LIST OF TABLES

Table 1	Apollo 10 Detailed Test Objectives
Table 2	Apollo 10 Achievements
Table 3	Apollo 10 Powered Flight Sequence of Events
Table 4	Apollo 10 Trans Lunar and Trans Earth Maneuver Summary
Table 5	Apollo 10 Lunar Orbit Maneuver Summary
Table 6	Apollo 10 Consumables Summary at End of Mission
Table 7	Apollo 10 Color TV Log
Table 8	Apollo 10 Space Vehicle Discrepancy Summary

Table 1. APOLLO 10 DETAILED TEST OBJECTIVES

LAUNCH VEHICLE

CATEG.	TITLE	ACCOMPLISHMENT SCHEDULE (DAYS) -							
		1	2	3	4	5	6	7	8
S	J-2 Engine Modification	X							
S	Engine environ. in S-II and S-IVB	X							
S	LV Longitudinal Oscillation environ/S-IC	X							
S	S-IC Mod suppression of low freq. osc.	X							
S	S-II longitudinal oscillation environ.	X							
S	S-II early CECO oscillation suppression	X							

SPACECRAFT

CATEG.	TITLE	ACCOMPLISHMENT SCHEDULE (DAYS)								
		1	2	3	4	5	6	7	8	
P20.78	CSM / LM Rendezvous Capability					X				
P16.10	LM Steerable Antenna Performance				X	X				
P20.121	Lunar Orbit Determination						X			
P20.91	Lunar Landing Site Determination				X					
P16.14	Landing Radar Test					X				
P20.66	Crew Activities Lunar Distance				X	X	X			
P11.15	PGNCS Undocked DPS Performance					X				
*S16.17	Relay modes Voice/TM				X	X			Partial	
S16.12	LM Omni Antennas Lunar Distance				X	X				
S16.15	Rendezvous Radar Performance					X				
S13-14	LM Supercritical Helium				X	X				
S12.9	Unmanned AGS Controlled APS Burn					X				
S20.77	VHF Ranging					X				

CATEG.	TITLE	ACCOMPLISHMENT SCHEDULE (DAYS)								
		1	2	3	4	5	6	7	8	
S20.86	Lunar Orbit Visibility				X	X				
S7. 26	Space Environment Thermal Control	X	X	X	X	X	X	X		
*S20.79	Passive Thermal Control Modes					X				Partial
S12.8	AGS CES Attitude/Translation Control					X				
S12.10	LM AGS Rendezvous Evaluation					X				
S20.82	PGNCS/AGS Monitoring				X	X				
S20.80	Ground Support Lunar Distance				X	X	X			
S13.13	Long Duration Unmanned APS Burn					X				
S20.117	LOI Maneuver					X				
S11.17	LM IMU Performance					X				
S 6.9	CSM High Gain Antenna Reflectivity							X		
S20.46	Transposition /Docking/LM Ejection	X								
S20.95	Midcourse Correction Capability		X							
S12.6	AGS Performance					X				
S 1.39	Midcourse Navigation/ Star-Lunar Landmark							X		
S20.83	LM Consumables Lunar Orbit				X	X				

<u>Other Major Activities Not Listed As DTO's:</u>

Color TV

Scientific and Engineering Photography

CSM S-Band Auto Reac. Test
LM Post APS Depl. Tests
(Comm and PGNS/AGS Switch)
Star/Earth Horiz Navig Sightings

<u>SUMMARY</u>

LV DTO's Secondary   6
SC DTO's
    Principal   7
    Secondary   22
All 100 pct except 2 second. SC DTO'S*

<u>Table 2. APOLLO 10 ACHIEVEMENTS</u>

1. SUCCESSFUL ACCOMPLISHMENT OF ALL PRIMARY OBJECTIVES

2. FIFTH SATURN V ON-TIME LAUNCH.

3. LARGEST PAYLOAD EVER PLACED IN EARTH ORBIT.

4. LARGEST PAYLOAD EVER PLACED IN LUNAR ORBIT.

5. FIRST DEMONSTRATION OF LUNAR ORBIT RENDEZVOUS.

6. FIRST DOCKED CSM/ LM LUNAR LANDMARK TRACKING

7. FIRST BURN OF DESCENT PROPULSION SYSTEM ENGINE IN THE LUNAR LANDING MISSION CONFIGURATION AND ENVIRONMENT.

8. FIRST EVALUATION OF THE LM STEERABLE ANTENNA AT LUNAR DISTANCES.

9. FIRST FIDELITY DEMONSTRATION OF LUNAR MISSION PROFILE (EXCEPT FOR PRE-DOI LANDMARK TRACKING, POWERED DESCENT, LUNAR SURFACE ACTIVITY AND ASCENT).

10. FIRST LOW LEVEL (50,000 FEET) EVALUATION OF LUNAR VISIBILITY.

11. FIRST EVALUATION OF THE LM OMNI DIRECTIONAL ANTENNAS AT LUNAR DISTANCE.

12. FIRST FLIGHT TEST OF THE ABORT GUIDANCE SYSTEM DURING A LONG DURATION ASCENT PROPULSION SYSTEM BURN.

13. FIRST INFLIGHT USE OF VHF RANGING.

14. FIRST LANDING RADAR TEST IN NEAR LUNAR ENVIRONMENT.

15. FIRST TIME DEMONSTRATION TRANSLUNAR MIDCOURSE CORRECTION CAPABILITIES WITH A DOCKED CSM/ LM.

16. FIRST DEMONSTRATION OF WESTINGHOUSE COLOR TV CAMERA IN FLIGHT.

17. FIRST MANNED NAVIGATIONAL VISUAL AND PHOTOGRAPHIC EVALUATION OF LUNAR LANDING SITES 2 AND 3.

18. FIRST MANNED VISUAL AND PHOTOGRAPHIC EVALUATION OF RANGE OF POSSIBLE LANDING SITES IN APOLLO BELT HIGHLANDS AREAS.

19. ACQUISITION OF MAJOR QUANTITIES OF PHOTOGRAPHIC TRAINING MATERIALS FOR APOLLO 11 AND SUBSEQUENT MISSIONS.

20. ACQUISITION OF NUMEROUS VISUAL OBSERVATIONS AND PHOTOGRAPHS OF SCIENTIFIC SIGNIFICANCE.

Table 3. APOLLO 10 MAY 26, 1969 POWERED FLIGHT SEQUENCE OF EVENTS

EVENT	PRELAUNCH PLANNED (GET) HR:MIN:SEC	ACTUAL (GET) HR: MIN : SEC
Range Zero (12:49:00.0 EDT)	00:00:00.0	00:03:00.0
Liftoff Signal (TB-1)	00:00:00.6	00:00:00.6
Pitch and Roll Start	00:00:12.5	00:00:12.5
Roll Complete	00:00:31.2	00:00:31.2
SIC Inboard Engine Cutoff (TB-2)	00:02:15.3	00:02:15.2
Begin Tilt Arrest	00:02:36.7	00:02:37.3
SIC Outboard Engine Cutoff (TB-3)	00:02:40.2	00:02:41.6
S-IC/S-II Separation	00:02:40.9	00:02:42.3
S-II Ignition	00:02:41.6	00:02:43.1
S-II Second Plane Separation	00:03:10.9	00:03:12.3
Launch Escape Tower Jettison	00:03:16.4	00:03:17.8
S-II Center Engine Cutoff	00:07:39.2	00:07:40.6
S-II Outboard Engine Cutoff (TB-4)	00:09:14.1	00:09:12.6
S-II/S-IVB Separation	00:09:15.0	00:09:13.5
S-IVB Ignition	00:09:15.1	00:09:13.6
S-IVB Cutoff (TB-5)	00:11:43.5	00:11:43.8
Earth Parking, Orbit Insertion	00:11:53.5	00:11:53.8
Begin S-IVB Restart Preparations (TB-6)	02:23:46.9	02:25:47.7
Second S-IVB Ignition	02:33:26.9	02:33:27.7
Second S-IVB Cutoff (TB-7)	02:38:48.6	02:38:49.5
Translunar Injection	02:38:58.6	02:38:59.5

*Prelaunch planned times are based on MSFC Saturn V AS-505 Apollo 10 mission LV operational trajectory dated April 17,1969 as revised by MSFC memo S and E AERO-FMT-106-69, May 5 1969.

Table 4. APOLLO 10 TRANSLUNAR AND TRANSEARTH MANEUVER SUMMARY

MANEUVER	GROUND TIME (GET) AT IGNITION (hr:min:sec)			BURN TIME (seconds)			VELOCITY CHANGE (feet per second - fps)			GET OF CLOSEST APPROACH HT. (NM) CLOSEST APPROACH		
	PRE-LAUNCH PLAN	REAL-TIME PLAN	ACTUAL	PRLAU PLAN	RLTIM PLAN	ACTUAL	PRE-LAU PLAN	RL-TIME PLAN	ACTUAL	PRE-LAU PLAN	RL-TIME PLAN	ACTUAL
TLI S-IVB IVB)	02:33:25.5	02:33:25.1	02:33:25.1	343.8	343.9	344.9	10429.4	10437.6	10437.0	76:13:22 957.0	76:10:15 907.5	76:22:02 712.6
Evasive Maneuver (SPS)	04:38:47.6	04:39:09.0	04:39.09.0	2.8	2.65	2.5	19.7	19.7	18.7	76:45:01.4 303.5	76:41:53 252.3	76:40:02 311.6
MCC-1 (SPS)	11:38:46.4	11:30:00	N.P.	8.1	6.6	N.P.	57.0	47.2	N.P.	75:49:40 58.4	75:52:00 59.0	N.P. N.P.
MCC-2(SPS)	26:33:00	26:32 :56.1	26:32:56.1	0	6.67	6.70	0	48.7	48.9	N.A. N.A.	76:00:19 58.7	76:00:11 65.9
MCC-3	53:45:00	53:45:00	N.P.	0	N.A.	N.P.	0	0.7	N.P.	N.A. N.A.	N.A. N.A.	N.P. N.P.
MCC-4	70:45:00	N.A.	N.P.	0	N.A.	N.P.	0	N.A.	N.P.	N.A. N.A.	N.A. N.A.	N.P. N.P.
Lunar Orbit Maneuvers	Lunar orbit -maneuvers are summarized on a separate table.									GET ENTRY INTERFACE (EI) VELOCITY (fps) AT EI FLIGHT PATH ANGLE AT EI		
TEI (SPS)	137:20:22.4	137:36:28	137:36:28	168.9	161.3	164	3622.5	3630.3	3625	191:50:32 36,309 -6.52	191:50:16 36,314 -6.52	191:48:46 36,315 -6.9
MCC -5	152:20:22.4	151:59:59	N.P.	0	1.8	N.P.	0	0.4	N.P	N.A. N.A. N.A.	191:50:11 36,314 -6.52	N.P. N.P. N.P.
MCC-6	176:50:32	176:49:58.4	N.P.	0	2.9	N.P.	0	1.3	N.P.	N.A N.A. N.A.	191:48:56 36,315 -6.52	N.P. N.P. N.P.
MCC-7	188:50:32	188:49:56.8	188:49:56.8	0	6.5	6.54	0	1.6	1.6	N.A. N.A. N.A.	191:48:54 36,315 -6.52	191:48:54 36,315 -6.53

Table 5. Date: May 26, 1969 APOLLO 10 LUNAR ORBIT MANEUVER SUMMARY

MANEUVER	GROUND ELAPSED TIME AT IGNITION (hr:min:sec) (GET)			BURN TIME (seconds)			VELOCITY CHANGE (feet per second - fps)			APOCYNTHION /PERICYNTHION RESULTANT (NAUTICAL MILES)		
	PR-LAU PLAN	REAL-TIME PLAN	ACTUAL	PRLAU PLAN	RL-TIM PLAN	ACTUAL	PR-LAU PLAN	RL-TIME PLAN	ACTUAL	PR-LAU PLAN	RL-TIME PLAN	ACTUAL
LOI-1(SPS)	75:45:43.2	75:55:53.3	75:55:53	362	353.9	356	2974	2981.5	2981.4	170/60	170.7/59.7	170.4/59.6
LOI-2 (SPS)	80:10:45.5	80:25:07.4	80:25:07	14.4	14	13.9	138.5	138.9	138.1	60/60	60.1/60.1	61.5/58.9
UNDOCK(SM RCS)	98:05:15.6	98:22:00	98:22:00	N.A.	N.A.	N.A.	N.A.	N.A.	N.A.	N.A.	N.A.	N.A.
CSM SEP (SM RCS)	98:35:15.6	98:47:16.0	98:47:16.0	6.9	8.1	10.4	2.5	2.5	3.2	59.2/60.1	62.1/57.9	61.9/58.0
DOI (LM DPS)	99:33:57	99:46:00.9	99:46:00.9	27.7	27.4	27.4	71.1	71.3	71.2	59.5/8.8	61.2/8.4	61.2/8.4
PHASING (LM DPS)	100:46:21	100:58:25.2	100:58:25.2	45.3	40.1	40.1	195.6	176.9	176.7	195.1/9.2	189.8/11.7	190.0/11.9
STAGING (LM RCS)	102:33:18	1102:45:00	102:45.00	0.0	0.0	0.0	-2.0/+2.0	0.0	0.0	195.1/9.2	189.8/11.7	190.1/11.8
INSERTION (LM APS)	102:43:18	102:55:01.4	102:55:01.4	15.4	15.3	15.5	206.9	220.9	221.0	45.8/8.6	45.8/11.1	45.3/11.2
CSI (LM RCS)	103:33:46	103:45:54.6	103:45:54.6	32.2	27.3	27.3	50.5	45.3	45.3	44.9/44.3	47.2/41.8	47.2/41.8
CDH (LM RCS)	104:31:43	104:43:52.0	104:43:52.0	2.3	3.4	3.7	3.4	2.8	3.1	44.4/44.2	47.6/42.2	46.8/42.1
TPI (LM RCS)	105:08:57	105:22: 55.0	105:22:55	16.1	14.7	28.8	25.4	24.3	24.0	61.8/43.8	58.1/46.8	58.0/46.8
DOCKING (SM RCS)	106:15:00	106:22:00	106:22:00	N.A.	N.A.	N.A.	N.A.	N.A.	N.A.	N.A.	N.A.	N.A.
P J SEP (SM RCS)	108:09:24	108:43:30.0	108:43:30	5.5	6.5	6.9	2.0	2.0	2.1	60.0/59.3	63.4/56.2	63.2/55.0
APS DEPL (LM APS)	108:38:57	108:51:01.0	108:51:01.0	214.5	246.5	212.93	3836.0	4600	3838.4	59.0	59.0	59.4

N.A. - Not Available

Table 6. DATE: May 26, 1969 APOLLO 10 CONSUMABLES-SUMMARY AT END OF MISSION

CONSUMABLE		LAUNCH LOAD	PRELAUNCH PLANNED REMAINING	ACTUAL REMAINING
CM RCS PROP (POUNDS /PERCENT)	U	207/100	107/51.7	Not Available
SM RCS PROP (POUNDS /PERCENT)	U	1,220/100	370/30.2	623/51.1
SPS PROP (POUNDS /PERCENT)	TK	40,590/100	4,154/10.2	3,336/8.2
SM HYDROGEN (POUNDS /PERCENT)	U	52.5/100	16.6/31.7	13.7/26.1
SM OXYGEN (POUNDS /PERCENT)	U	615.6/100	263.6/42.8	267.7/43.5
LM RCS PROP (POUNDS /PERCENT)	U	548.9/100	*134/24.4	*77/14.0
LM DPS PROP (POUNDS /PERCENT)	U	17,741/100	**16,893/95.4	**16,982/95.9
LM APS PROP (POUNDS /PERCENT)	U	2,567/100	*0/0	*0/0
LM A/S OXYGEN (POUNDS /PERCENT)	T	4.72/100	*3.37/71.4	*4.08/86.5
LM D/S OXYGEN (POUNDS /PERCENT)	T	48.7/100	**42.2/86.6	**42.4/87.0
LM A/S WATER (POUNDS /PERCENT)	T	84.8/100	*32.5/61.9	*47.5/56.0
LM D/S WATER (POUNDS /PERCENT)	T	318.1/100	**257.8/81.1	**262.7/82.6
LM A/S BATTERIES (AMP-HRS /PERCENT)	T	592/100	*282/47.6	*274.9/46.4
LM D/S BATTERIES (AMP-HRS /PERCENT)	T	1,600/100	**1150/71.9	*1160.2/72.5

U — Usable quantity
TK — Tank quantity
T — Total quantity

*At termination of APS burn to depletion
**At descent stage jettison

Table 7. APOLLO 10 COLOR TV LOG

Number		Transmission Start		Transmission Time (Min/Sec)			
Plan	Actual	Planned	Actual	Planned	Actual	STA	EVENT
1	1	3:03:24	3:06:00	15:00	22:00	GDS	SEPARATION, TRANSPOSITION and DOCKING
	2		3:56:00		13:25	GDS	CSM/LM EJECTION FROM S-IVB
	3		5:06:34		13:15	GDS	VIEW OF EARTH and S/C INTERIOR
	4		7:11:27		24:00	GDS	VIEW OF EARTH and S/C INTERIOR
2	5	27:15:00	27:00:48	15:00	27:43	GDS	VIEW OF EARTH and S/C INTERIOR
	6		48:00:51		14:39	MAD	VIEW OF EARTH and S/C INTERIOR (RECORDED)
	7		48:24:00		3:51	MAD	VIEW OF EARTH and S/C INTERIOR (RECORDED)
	8		49:54:00		4:49	GDS	VIEW OF EARTH
3	9	54:00:00	53:35:30	15:00	25:00	GDS	VIEW OF EARTH and S/C INTERIOR
4	10	72:20:00	72:37:26	15:00	17:16	GDS	VIEW OF EARTH and S/C INTERIOR (BOTH 210' and 85' DISH)
5	11	80:45:00	80:44:40	10:00	29:09	GDS	VIEW OF LUNAR SURFACE
6	12	98:13:00	98:29:20	10:00	20:10	GDS	VIEW OF SEPARATION MANEUVER
7		108:35:50		15:00		GDS	
8		126:20:00		40:00		GDS	
	13		132:07:12		24:12	GDS	VIEW OF LUNAR SURFACE and S/C INTERIOR
9	14	137:45:00	137:50:51	15:00	43:03	HSK	VIEW OF MOON POST TEI
	15		139:30:16		6:55	HSK	VIEW OF MOON POST TEI
	16		147:23:00		11:25	GDS	VIEW OF RECEDING MOON and S/C INTERIOR
10	17	152:35:00	152:29:19	10:00	29:05	GDS	VIEW OF EARTH, MOON and S/C INTERIOR
	18		173:27:17		10:22	GDS	VIEW OF EARTH and S/C INTERIOR
11	19	186:50:00	186:51:49	15:00	11:53	GDS	VIEW OF EARTH and S/C INTERIOR
Total				2:55:00	5:52:12		

Quality, color and resolution of color TV was outstanding. Resolution remained excellent throughout focussing range of 6:1 zoom lens.

Table 8. APOLLO 10 SPACE VEHICLE DISCREPANCY SUMMARY

LAUNCH VEHICLE DISCREPANCY SUMMARY

S-IVB STAGE AUXILIARY HYDRAULIC PUMP MOTOR AMPERAGE DROPPED FROM NORMAL READING ABOUT 220 SECONDS INTO S-IVB SECOND BURN. THE HYDRAULIC SYSTEM BURN ALSO INDICATED MARGINAL PUMP PERFORMANCE.

COMMAND/ SERVICE MODULE DISCREPANCY SUMMARY

* EDS MODULE LIGHT BULBS FAILED INTERMITTENTLY. (PRELAUNCH)

* COMMAND MODULE REACTION CONTROL SYSTEM A HELIUM MANIFOLD PRESSURE DECAY. (PRELAUNCH)

* FUEL CELL 1 OXYGEN FLOWMETER FAILED. (PRELAUNCH)

* COMMAND MODULE REACTION CONTROL SYSTEM B HELIUM MANIFOLD PRESSURE ABRUPTLY DROPPED FROM 44 TO 37 PSI WHEN PROPELLANT ISOLATION VALVES OPENED. (PRELAUNCH)

* SUIT LOOP WATER SEPARATOR BREAKTHROUGH; CHANGE IN WICK WETTING TECHNIQUE WAS SUCCESSFUL.

* PRIMARY ENVIRONMENTAL CONTROL SYSTEM EVAPORATOR DRIED OUT. SWITCHED TO SECONDARY EVAPORATOR AT 0:15:00. PRIMARY EVAPORATOR RESERVICED AT 73:15:00, DRIED OUT AGAIN.

* THIN WHITE LINE ON RIGHT HAND SIDE WINDOW (TOP TO BOTTOM).

* CREW REPORTED HIGH FREQUENCY VIBRATION PRIOR TO COMPLETION OF S-IVB TRANSLUNAR INJECTION FIRING.

* HIGH OXYGEN FLOW CAUTION AND WARNING DURING TRANSLUNAR INJECTION FIRING. CABIN OXYGEN REGULATORS OPERATED AT SAME TIME.

* CARBON DIOXIDE PARTIAL PRESSURE READINGS 1.2 MM HG; SHOULD BE LOWER.

* PROGRAM ALARM 122 OCCURRED WHILE OR IMMEDIATELY AFTER THE CREW WAS OBSERVING THE EARTH THROUGH OPTICS.

* ENVIRONMENTAL CONTROL SYSTEM OXYGEN MANIFOLD PRESSURE DROPPED TO 75 PSI (SHOULD BE 100) FOR ABOUT 3 SECONDS DURING REDUNDANCY CHECKS.

* WATER PROBLEMS: WATER /GAS SEPARATOR DID NOT OPERATE SATISFACTORILY. AIR IN INITIAL SERVICED POTABLE WATER.

* DIGITAL EVENT TIMER ON PANEL 1 JUMPED 2 MINUTES WHILE COUNTING DOWN FIRST MIDCOURSE CORRECTION

* THERMAL COATING ON FORWARD HATCH FLAKED OFF DURING LUNAR MODULE CABIN PRESSURIZATION.

* TUNNEL WOULD NOT VENT.

* SIMPLEX-A NOT OPERATING; OPERATED PROPERLY LATER.

* NO DOWN-VOICE FROM COMMAND AND SERVICE MODULE.

* TRANSPONDER IN COMMAND MODULE HAD TO BE CYCLED TO GET RENDEZVOUS RADAR TRACKING.

* LUNAR MODULE CAMERA FAILURES.

* AC CIRCUIT BREAKER TO FUEL CELL 1 OPEN; RESET GAVE UNDERVOLTAGE INDICATION.
FUEL CELL OPEN-CIRCUITED AT 121:16. SHORT IN GLYCOL PUMP, HYDROGEN PUMP , PH SENSOR, OR ASSOCIATED WIRING.

* CONDENSER EXIT TEMPERATURE ON FUEL CELL 2 FLUCTUATING AND TRIGGERED CAUTION AND WARNING SEVERAL TIMES.

* CRYOGENIC HYDROGEN AUTOMATIC HEATER CONTROL DID NOT TURN OFF (SIMILAR OCCURRENCE ON APOLLO 9 WITH CRYOGENIC OXYGEN HEATER).

* FUEL CELL 1 PURGE VALVE DID NOT CLOSE UNTIL SWITCH CYCLED AT WHICH TIME FLOW SLOWLY DECREASED TO 0.04 LB/HR IN 40 MINUTES.

LUNAR MODULE DISCREPANCY SUMMARY

* CREW REPORTED THAT THE LUNAR MODULE POTABLE WATER CONTAINED AIR THROUGH USE.

* NOISE IN CABIN

* S-BAND ANTENNA MOVEMENT NOISE.

* GLYCOL PUMP NOISE BAD.

* FANS

* ABORT GUIDANCE SYSTEM DEADBAND SWITCH INDICATED MAX. ON TELEMETRY WHEN IN MIN. POSITION.

LUNAR MODULE INSTRUMENTATION DISCREPANCIES

* GLYCOL, TEMPERATURE READ ZERO DURING FIRST MANNING; (82:45:00)

* CHAMBER PRESSURE SWITCHES FAILED ON REACTION CONTROL THRUSTERS B3D (97:34), B4U (98:31).

* CASK TEMPERATURE READ PROPERLY PRELAUNCH, READ OPEN INFLIGHT.

* REACTION CONTROL SYSTEM A MANIFOLD PRESSURE WENT TO ZERO (108:30).

* DESCENT PROPULSION FUEL PRESSURE GQ3501 READ ZERO IN CABIN TELEMETRY NORMAL.

* ASCENT PROPULSION OXIDIZER MANIFOLD PRESSURE 187 ON TELEMETRY, 180 IN CABIN; PREDICTED WAS 170 PSI (GP1503).

* ABSOLUTE PRESSURE IN GLYCOL SYSTEM NOT STEADY.

* SIMPLEX-A NOT OPERATING AT 94:28:00.

* BACKUP VOICE NOISY BUT READABLE WHILE ON OMNI IN REVOLUTION 13.

* S-BAND STEERABLE ANTENNA OPERATION SHOWED DROP IN SIGNAL DURING PART OF REVOLUTION 13. ATTITUDE'S WERE PROPER FOR MAINTAINING LOCK.

* RENDEZVOUS RADAR ALARMS DURING FIRST MARKS, INDICATING BAD INPUTS TO COMPUTER.

101:09:29 - RANGE RATE READ MINUS 9800 FT/SEC; SHOULD BE ABOUT 200 FT/ SEC

103:14:24 - READ 9999 FT/SEC; SHOULD BE 295 FT/SEC

104:37:00 - READ 6722 FT/SEC; SHOULD BE MINUS 124 FT/SEC (NO ALARM).

105:17:00 - RANGE READ 22 N. MI.; SHOULD BE 40 N. MI. (RANGE RATE MAY ALSO HAVE BEEN IN ERROR).

* ABORT GUIDANCE VERSUS VERB 83 IN PRIMARY GUIDANCE COMPUTER LOCAL VERTICAL DIFFERENT BY 20 DEG; NO DIFFERENCE AFTER UNDOCK.

* APPARENT GIMBAL DRIVE ACTUATOR PITCH FAILURE; TELEMETRY INDICATES ACTUATOR NEVER MOVED.

* AC TRANSIENT MAY BE ASSOCIATED; 124 V PEAK (PHASING BURN).

* CAUTION AND WARNING ON ACTUATOR FAIL DURING PHASING BURN.

* LARGE ATTITUDE EXCURSIONS DURING STAGING. ABORT GUIDANCE MODE CONTROL SWITCH WAS IN "AUTO" RATHER THAN "ATTITUDE HOLD" AND APPARENTLY IN MAXIMUM DEADBAND.

* MANUAL CONTROL OVERRODE. THE ABORT GUIDANCE "AUTO" CONFIGURATION WAS ATTEMPTING.

* TO POINT THE LUNAR MODULE AXIS AT THE COMMAND MODULE AS SET IN TO THE; ABORT GUIDANCE. SIMULATION RESULTS SHOW WHAT HAPPENED IS ASSOCIATED WITH THE ABORT GUIDANCE MODE SWITCH.

* DURING DESCENT PROPULSION PHASING, CAUTION AND WARNING AND ALARM.

* GIMBAL WARNING.

* LOW LEVEL SENSOR.

* CREW COMMENTED THAT RETICLE WAS BAD DURING PLATFORM REALIGNMENT AFTER PHASING BURN.

* GLYCOL PUMP SWITCHOVER PRIOR TO ASCENT PROPULSION FIRING DURING CABIN CLOSEOUT.

* INVESTIGATE 3.5 DEG ALIGNMENT DIFFERENCE BETWEEN LUNAR MODULE AND COMMAND MODULE PRIOR TO SEPARATION AT 96:30.

* ABNORMAL RISE IN CARBON DIOXIDE INDICATION ON PRIMARY LIOH CARTRIDGE.

* LUNAR MODULE CABIN PRESSURE DROPPED ABRUPTLY AT COMMAND MODULE/LUNAR MODULE SEPARATION.

* RATE ERROR NEEDLES INDICATION OFF BY 0.2 DEG IN PITCH, 0.3 DEG IN YAW, AND 0.2 DEG IN ROLL.

POST LAUNCH MISSION OPERATION REPORT NO.2

Report No. M-932-69-10
OFFICE OF MANNED SPACE FLIGHT

TO: A/Administrator 18 July 1969

FROM: MA/Apollo Program Director

SUBJECT: Apollo 10 Mission (AS-505) Postlaunch Mission Operation Report #2

Review of the Apollo 10 Mission results since issuance of Postlaunch Mission Operation Report No. 1 (26 May 1969) indicates that all mission objectives were attained. Further detailed analysis of all data is continuing and appropriate refined results of the mission will be reported in Manned Space Flight Center technical reports.

Attached is Postlaunch Mission Operation Report No. 2, which updates or supplements Report No. 1 and includes our assessment of the mission. Based on the mission performance as described in these reports, it is recommended that the Apollo 10 Mission be adjudged as having achieved agency preset primary objectives and be considered a success.

Sam C. Phillips
Lt. General, USAF
Apollo Program Director

APPROVAL:

George E. Mueller

Associate Administrator for Manned Space Flight

Prepared by: Apollo Program Office-MAO

FOR INTERNAL USE ONLY

<u>NASA OMSF PRIMARY MISSION OBJECTIVES FOR APOLLO 10</u>

<u>PRIMARY OBJECTIVES</u>

Demonstrate crew/space vehicle/mission support facilities performance during a manned lunar mission with CSM and LM.

Evaluate LM performance in the cislunar and lunar environment.

Sam C. Phillips   George E. Mueller
Lt. General, USAF   Associate Administrator for
Apollo Program Director   Manned Space Flight

Date: 6 May 1969   Date: May 8 1969

<u>RESULTS OF APOLLO 10 MISSION</u>

Based upon a review of the assessed performance of Apollo 10, launched 18 May 1969 and completed 26 May 1969, this mission is adjudged a success in accordance with the objectives stated above.

Sam C. Phillips   George E. Mueller
Lt. General, USAF   Associate Administrator for
Apollo Program Director   Manned Space Flight

Date: 18 July 1969

<u>GENERAL</u>

As stated in Postlaunch Mission Operation Report No. 1, all elements of the Apollo system performed satisfactorily during the Apollo 10 Mission. Tables 1 and 2 provide updated values for the mission sequence of events. Tables 3 through 6 present summaries of the major discrepancies experienced by the launch vehicle, Command/Service Module, Lunar Module, and mission support.

<u>Table 1. APOLLO 10 SEQUENCE OF EVENTS</u>

EVENT	GROUND ELAPSED TIME (GET) (hr: min: sec)	
	PLANNED*	ACTUAL
Range Zero (12:49:00.0 EDT)	0:00:00.0	0:00:00.0
Liftoff Signal (TB-1) (Time Base -1)	0:00:00.6	0:00:00.6
Begin Pitch and Roll	0:00:12.5	0:00:12.5
Roll Complete	0:00:31.2	0:00:31.2
Maximum Dynamic Pressure (Max Q)	0:01:17	0:01:22.6
S-IC Inboard Engine Cutoff (TB-2)	0:02:15.3	0:02:15.2
Begin Tilt Arrest	0:02:36.7	0:02:37.3
S-IC Outboard Engine Cutoff (TB-3)	0:02:40.2	0:02:41.6
S-IC/S-II Separation	0:02:40.9	0:02:42.3
S-II Ignition	0:02:41.6	0:02:43.1

EVENT	GROUND ELAPSED TIME (GET) (hr: min: sec)	
	PLANNED*	ACTUAL
Jettison S-II Aft Interstage	0:03:10.9	0:03:12.3
Launch Escape Tower Jettison	0:03:16.4	0:03:17.8
S-II Center Engine Cutoff	0:07:39.2	0:07:40.6
S-II Outboard Engine cutoff (TB-4)	0:09:14.1	0:09:12.6
S-II/S-IVB Separation	0:09:15.0	0:09:13.5
S-IVB Ignition	0:09:15.1	0:09:13.6
S-IVB Cutoff (TB-5)	0:11:43.5	0:11:43.8
Earth Parking Orbit Insertion	0:11:53.5	0:11:53.8
Begin S-IVB Restart Preparations (TB-6)	2:23:46.9	2:23:47.7
Second S-IVB Ignition (Translunar Injection)	2:33:26.9	2:33:27.7
Second S-IVB Cutoff (TB-7)	2:38:48.6	2:38:49.5
CSM/S-IVB Separation, SLA Panel Jettison	3:00:00	3:02:51
CSM Transposition and Docking Complete	3:10:00	3:17:38
CSM/LM Ejection from S-IVB/IU/SLA	4:09:00	3:56:24
CSM/LM SPS Evasive Maneuver	4:29:00	4:39:09.8
S-IVB Slingshot Maneuver (Propellant Dump)	4:39:00	4:49:00
Midcourse Correction No. 1 (MCC-1)	11:33:00	Not Required
Midcourse Correction No. 2 (MCC-2)	26:33:00	26:32:56.8
Midcourse Correction No. 3 (MCC-3)	53:45:00	Not Required
Midcourse Correction No. 4 (MCC-4)	70:45:00	Not Required
Lunar Orbit Insertion No. 1**	75:45:00	75:55:54
Lunar Orbit Insertion No. 2 (Circularization)	80:10:00	80:25:07
Intravehicular Transfer to LM	94:25:00	95:02:00
Undocking	98:05:00	98:22:00
LM/CSM Separation Maneuver	98:35:00	98:47:16
Descent Orbit Insertion	99:34:00	99:46:01
Phasing Maneuver	100:46:00	100:58:25
LM Descent Stage Jettison	102:33:00	102:45:00
Ascent Insertion Maneuver	102:43:00	102:55:01
Concentric Sequence Initiation	103:34:00	103:45:55
Constant Delta Height Maneuver	104:32:00	104:43:52
Terminal Phase Initiation	105:09:00	105:22:55
Docking	106:20:00	106:22:08
LM Ascent Stage Jettison	108:09:00	108:24:37
APS Burn to Propellant Depletion	108:39:00	108:51:01
Transearth Injection (Ignition)	137:20:00	137:36:28
Midcourse Correction No. 5 (MCC-5)	152:20:00	Not Required
Midcourse Correction No. 6 (MCC-6)	176:50:00	Not Required
Midcourse Correction No. 7 (MCC-7)	188:50:00	188:49:57
CM/SM Separation	191:35:00	191:33:24
Entry Interface (400,000 feet altitude)	191:50:00	191:48:54
Enter S-band Blackout	191:50:26	191:49:21
Exit S-band Blackout	191:53:26	191:52:22
Drogue Parachute Deployment	191:58:33	191:57:11
Main Parachute Deployment	191:59:22	191:58:01
Landing	192:04:00	192:03:23

*Prelaunch planned times are based on MSFC Saturn V AS-505 Apollo 10 Mission IV Operational Trajectory, April 17, 1969, as revised by MSFC memo no. S and E-AERO-FMT-106-69, May 5, 1969; and on MSC Revision 1 of Spacecraft Operational Trajectory for Apollo 10, April 28, 1969.

**Delay of the first midcourse correction to the MCC-2 option caused the translunar trajectory to be longer than planned thus delaying lunar orbit events approximately 12 minutes.

Table 2. APOLLO 10 RECOVERY EVENTS

EVENT	EDT
First visual contact	12:40
First radar contact	12:41
Visual contact by USS PRINCETON	12:45
First voice contact	12:50
Landing (164°41' W.Long., 15°1' S.Lat.)	12:52
Flotation collar installed	1:10
CM hatch open	1:17
Crew in raft	1:20
Crew aboard helicopter	1:26
Crew aboard USS PRINCETON	1:31
CM aboard USS PRINCETON	2:22

Table 3. LAUNCH VEHICLE DISCREPANCY SUMMARY

DESCRIPTION	REMARKS
The S-IVB auxiliary hydraulic pump stopped producing full pressure during the second burn.	Probably caused by structural failure of the compensator spring guide. The spring guide was replaced on AS-506 (Apollo 11).
At T-9 hours, the ECS air/nitrogen purge duct in the IU failed, apparently at the duct joint 4 inches inside the stage skin.	A second clamp will be added over the duct at the joint and the screw torque on both clamps will be increased. This change has been completed on AS-506.
Low frequency (19-Hz) longitudinal and lateral oscillations were present during both S-IVB burns. High-frequency (46-Hz) oscillations occurred during the latter part of the second S-IVB burn.	The 19-Hz vibration is a normal low-level response to the uncoupled J-2 engine thrust oscillations. The 46-Hz vibration is attributed to excitation of the forward skirt ring mode by noise originating in the liquid hydrogen non-propulsive vent system. This condition is not a constraint on Apollo 11.

Cernan and Apollo 11 lunar module pilot Edwin Aldrin.

Table 4. COMMAND/SERVICE MODULE DISCREPANCY SUMMARY

DESCRIPTION	REMARKS
CM Reaction Control System A developed a small leak in the helium manifold prior to launch.	On future missions, the RCS pressure will be monitored for approximately 1 month prior to launch to ensure early detection of any leaks.
Rupture of the RCS oxidizer burst disc was noticed when the CM Reaction Control System B helium manifold pressure abruptly dropped from 44 to 37 psi as the propellant isolation valves were opened.	There is no mission impact as long as the shutoff valves and the engine control valves hold leak-tight. A burst disc leak check has been added after RCS propellant servicing.
The primary ECS evaporator dried out during the launch phase and again during the second lunar revolution.	A check of the water control microswitch assembly revealed that the actuator travel was at times not sufficient to open the switch. Actuator rigging procedures will be modified to assure proper overtravel.
Water problems on the flight were: (1) the water/gas separator did not operate satisfactorily; (2) air was contained in the ground serviced potable water.	Three new designs for the water/gas separator have been tested with good results. Operational procedures are being prepared for possible use on Apollo 11.
For about 2 hours on the seventh day of the flight, the flow from the CM water dispenser appeared to be less than normal.	The gun was probably clogged by excess O-ring lubricant. Should the gun become clogged in flight, several alternate sources are available for drinking water.
The thermal blanket on the CM forward hatch flaked off during LM cabin pressurization. Particles went throughout both spacecraft, requiring clean-up and causing crew discomfort.	The thermal blanket has been deleted effective on CSM 107 (Apollo 11).
The tunnel would not vent when the crew tried to perform the hatch integrity check prior to undocking. An alternate venting procedure through the LM had to be used.	Due to a ground procedural error, the vent line was terminated by a plug instead of a fitting with holes in it. CSM 107 and 108 have been verified as being properly configured.
Twice during lunar revolution 10, transmissions from the LM on VHF Simplex-A were not received in the CM. VHF Simplex-A operated satisfactorily for both voice and ranging during the remainder of the mission.	The most probable cause for the apparent failure of VHF Simplex-A was that, because of the numerous switch configuration changes in both vehicles, the LM & CSM were not configured simultaneously for communications on Simplex-A.
The CM rendezvous radar transponder had to be cycled to obtain operation following undocking.	Postflight testing of the switch and wiring did not reveal any problem. Failure analysis of the switch is continuing.
The CM 16mm sequence camera stopped operating in the pulse mode at 173:00 GET.	The problem has been traced to a faulty microswitch. All cameras on Apollo 11 will have high-reliability microswitches.
The fuel cell 1 AC circuit breaker tripped at 120:47 GET, due to a short circuit in the AC pump package.	The most probable cause was a breakdown in the insulation within the hydrogen pump.

DESCRIPTION

Condenser exit temperature of fuel cell 2 fluctuated and triggered the caution and warning light several times.

During the extended purge of fuel cell 1, the cryogenic hydrogen automatic pressure control twice failed to turn off the hydrogen tank heaters. The heaters were controlled manually after 170:30 GET.

When the purge valve was closed following the extended purge of fuel cell 1, it took 30 minutes for the hydrogen flow to decrease to zero, and an overpressure of 9 psi occurred in the regulated hydrogen supply to the fuel cell.

Four of the ten light bulbs in the launch vehicle engine warning annunciator failed intermittently prior to launch.

The stylus of the entry monitor system stopped scribing while initializing after the pre-entry tests. The scribe worked after slewing the scroll back and forth.

The VHF beacon antenna did not deploy on entry; however, three helicopters received the beacon signals.

Two retaining springs for the tunnel shaped charge holder ring did not capture on the minus y side.

The data storage equipment tape recorder slowed down several times during entry.

The stabilizer, which maintains the couch position when the foot strut of the center couch is removed, was connected during launch.

The gyro display coupler drifted excessively in roll and yaw (5° in 20 minutes) after performing properly early in the mission.

During one injection, the chlorine ampoule leaked, and no water sample could be drawn.

REMARKS

The cause has not yet been determined. This behavior is not detrimental to fuel cell component life or performance, but it does represent a nuisance to the crew because the caution and warning light must be reset manually.

If long-duration purges are required in the future, the heaters will be operated manually The Apollo Operations Handbook has been changed appropriately.

Extended hydrogen purge should not be conducted if preheat capability is lost, because of the effect of low temperatures on the hydrogen regulators (valve seat warpage and leakage). The Apollo Operations Handbook will be changed to caution against such an operation.

Caused by cold solder joints where the lamp leads are attached to the printed circuit board. The units for Apollo 11 and subsequent vehicles have been screened.

Subsequent to Apollo 12, either the scroll emulsion base will be made using the originally formulated soap or pressure-sensitive paper which was recently qualified will be used for the scroll.

The antenna did not deploy because one radial was caught under the outboard edge of the ramp. No change is required for Apollo 11.

Based on the Apollo 10 flight experience, ground tests, and analytical results, the probability of a failure to capture the charge holder under normal separation conditions is judged to be low.

The recorder vent valve allowed enough differential pressure to build up to deform the recorder cover sufficiently to contact the tape reel. For future missions, vent valves will be selected that open at the low end of the allowable range in the specification.

The stabilizer must be in the stowed position to allow strut stroking during an abort landing. A mandatory inspection step has been added to the pre-ingress checklist.

Postflight tests are in progress.

Postflight inspection revealed no defects. Probably caused by improper insertion of the ampoule.

DESCRIPTION	REMARKS
Relative motion occurred between the CM and LM at the docking ring interface when the CM RCS roll thrusters were fired.	Pressure in the tunnel reduced the frictional force at the docking ring interface. Proper tunnel venting capability has been verified on CSM 107 and 108.
The left-hand x-x head strut lockout torsion spring was found on the wrong side of the retention pin during the postflight check of lockout lever forces.	Correct installation on CSM 107 and 108 will be verified at KSC.
The digital event timer on panel I advanced a total of 2 minutes during the countdown for the first midcourse correction. At other times, the tens of seconds failed to advance.	A screening test has been developed for the timers installed in future spacecraft; however, the capability of the test to isolate unreliable timers has not yet been proven.

Table 5. LUNAR MODULE DISCREPANCY SUMMARY

DESCRIPTION	REMARKS
A master alarm with an engine pitch gimbal fail warning occurred during the DPS phasing burn. Telemetry shows both actuators operated normally.	Apparently caused by gimbal coasting. For Apollo 11 and subsequent missions, the brake mechanism has been redesigned and the fail warning has been made less sensitive to coasting.
Two master alarms with DPS propellant low quantity warnings occurred during the phasing burn.	A gas bubble probably uncovered the low level sensor accompanied by an intermittent open circuit in the low level signal circuit. The system has been greatly simplified on Apollo 11 and subsequent.
During revolution 13 the signal at the Mission Control Center was very weak when the LM omni antenna was used with backup downvoice.	The problem occurred in equipment at the Goldstone station, and an investigation is being conducted by Goddard Space Flight Center.
During the beginning of revolution 13, the S-band steerable antenna did not track properly for about 13 minutes, but it tracked well both before and after this period.	The crew may have inadvertently switched the track-mode switch to OFF instead of to AUTO at the time S-band acquisition was established.
The LM drinking water contained gas.	Consideration is being given to placing a gas separator in the drinking line.
The cabin was excessively noisy, primarily because of the glycol pump. The cabin fans and S-band steerable antenna also were noisy.	Modification of the LM hardware does not appear practical. Ear muffs or ear plugs will be provided for the crew to use during sleep periods.
During checkout of the Commander's oxygen purge system, the heater light did not come on.	Manned tests have shown that the gas temperature is acceptable for comfort without the heater.
The output from the yaw rate gyro did not always correspond to the actual LM yaw rate from 50 seconds prior to staging until several seconds after staging, but the output was normal before and after this period.	Data are being analyzed to identify the characteristics of the abnormal operation and to determine the failure mode.

DESCRIPTION	REMARKS
The dump of the LM low-bit-rate PCM data recorded in the CM stopped abruptly at 99:38:52 GET instead of continuing until approximately 99:46:00 GET.	The CM was reconfigured from the voice and data mode to the ranging mode approximately 11 minutes early because the times on the flight plan were incorrect since lunar orbit was initiated 12 minutes later than originally planned.
When the LM Ascent Stage separated from the CM, the LM cabin pressure dropped from 4.86 psia to less than 1.0 psia in 0.3 second.	The residual tunnel pressure of 4.86 psia, plus the pressure generated when the separation pyrotechnics fired, caused the latch on the LM tunnel hatch to fail and the cabin then vented through the hatch.
The CO_2 level and rate of increase were abnormally high while using the primary lithium hydroxide cartridge. The level decreased rapidly to the predicted level when the secondary cartridge was selected.	Lithium hydroxide cartridge variations combined with CO_2 sensor tolerances could account for the observed flight performance. Predictions for future flights will use more realistic operational characteristics.
Large attitude excursions occurred prior to and during LM staging.	For an undetermined reason the abort guidance system mode control changed from "attitude-hold" to "automatic" coincident with each vehicle gyration.
During the low-altitude lunar pass the Hasselblad 70mm camera stopped because of film binding in the magazine.	The Apollo 11 cameras and magazines will be inspected for damage, clearances, and contamination. The 1.6-ampere fuses will be replaced with 1.2-ampere high-reliability fuses.
During the low-altitude lunar pass, the LM 16mm camera failed to operate with magazine F installed. Magazine F was reinstalled later and the camera operated satisfactorily.	Magazine F had marginal clearances at the interface surfaces and edges. All magazines for subsequent missions will be fit-checked before flight.
Three operational anomalies occurred during use of the LM optical system: contamination of the reticle of the alignment optical telescope by hair-like material, difficulty in operating the dimmer control rheostat of the computer control and reticle dimmer, and disappearance of stars at approximately six star diameters from the center of the reticle.	Contamination of the reticle may have occurred through a gap in the housing that is required to allow for thermal expansion. The operation of the dimmer control rheostat as described by the crew was normal. Disappearance of stars may have been caused by contamination on the prism; the LM-5 (Apollo 11) prism and reticle have been cleaned and inspected.

Table 6. MISSION SUPPORT DISCREPANCY SUMMARY

DESCRIPTION	REMARKS
During line chilldown in preparation for LOX loading at about T-8 hours, the primary LOX replenish pump failed to start.	A blown fuse was found in the pump motor starter circuit. Troubleshooting and fuse replacement delayed completion of LOX loading by 50 minutes, but the built-in 1-hour hold at T-3:30 hours prevented a launch delay.

The Apollo 10 and Apollo 11 crews. Around the table from left to right is Michael Collins, Edwin (Buzz) Aldrin, Eugene Cernan, Thomas Stafford, Neil Armstrong and John Young.

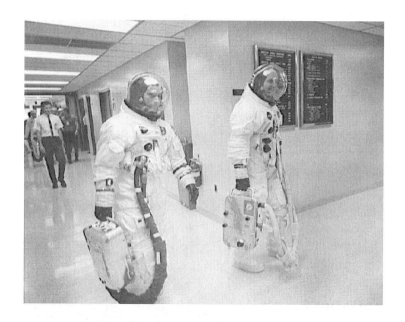

On their way to Launch Pad 39, John Young (Left) and Thomas Stafford (right) tote their portable air-conditioners to keep their suits cool.

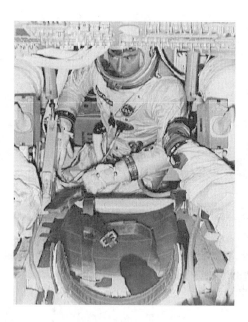

John Young during simulator training for the Apollo 10 flight.

One of the 1300 70mm images shot by Apollo 10. Most of the film
was used to take strings of pictures of the lunar surface. These
pictures were subsequently used by NASA to help pick future
landing sites.

Apollo 10 is attached to the Saturn V rocket in the Hi-Bay of the
Vehicle Assembly Building prior to launch.

Earthrise above the lunar surface. One of a sequence of seventeen
70mm images which were taken by Apollo 10 of the spectacular
event. When the images are run at high speed and sequentially they
provide a brief but dazzling 70mm movie.

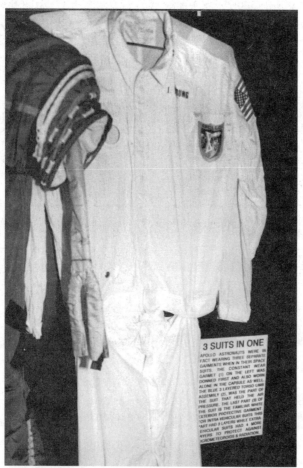

John Young's Constant Wear Garment from Apollo 10.
Now on display at the Michigan Space & Science Center in Jackson Michigan.

A Saturn V rocket is on display in Huntsville Alabama where it was designed by Wernher Von Braun. It is a sad but awe inspiring reminder of another era. It was probably to have been used for one of the cancelled lunar missions — Apollos 18 through 20.